A NOVEL DEFENSE

of SCIENTIFIC REALISM

A NOVEL DEFENSE

of SCIENTIFIC REALISM

Jarrett Leplin

New York   Oxford
Oxford University Press
1997

Oxford University Press

Oxford New York Athens Auckland Bangkok Bogota Bombay Buenos Aires
Calcutta Cape Town Dar es Salaam Dehli Florence Hong Kong
Istanbul Karachi Kuala Lumpur Madras Madrid Melbourne
Mexico City Nairobi Paris Singapore Taipei Tokyo Toronto

and associated companies in
Berlin Ibadan

Copyright © 1997 by Jarrett Leplin

Published by Oxford University Press, Inc.
198 Madison Avenue, New York, New York 10016

Oxford is a registered trademark of Oxford University Press

Library of Congress Cataloging-in-Publication Data
Leplin, Jarrett.
A novel defense of scientific realism / Jarrett Leplin.
p.   cm.
Includes bibliographical references and index.
ISBN 0-19-511363-2
1. Realism. 2. Science—Philosophy.   I. Title.
Q175.32.R42L46   1997
501—dc20   96-29084

9 8 7 6 5 4 3 2 1

Printed in the United States of America
on acid-free paper

To the memory of Emanuel Leplin,

a "natural born conductor,"

in the words of Pierre Monteux,

and a composer whose music,

in the words of Leonard Bernstein,

"is extraordinary, moving,

and should be heard."

Acknowledgments

I wish to thank William Newton-Smith, Peter Achinstein, and André Kukla for criticism. My greatest debt is to Larry Laudan, discussions with whom, over many years, have been invaluable to my efforts.

This project was begun with a summer research stipend from the National Endowment for the Humanities, and supported by the Research Council of the University of North Carolina, Greensboro.

Contents

Introduction xi

1 Truth and the Success of Science 3
   Explaining Scientific Success 3
   Simplistic Accounts of Success 6
   Dismissive Attitudes toward Explanation 9
   Underdetermination 12
   The Alleged Superfluousness of Truth in Explanation 15
   Excising Truth from Explanation 21
   Surrogates for Truth 28
   The Metaphysical Import of Truth 29

2 Conceptions of Novelty 34
   The History of the Independence Requirement 34
   Temporal Constraints 40
   Bayesian Confirmation 44
   Contemporary Analyses of Independence 49
   Constraints on the Analysis of Novelty 63

3   An Analysis of Novelty   64

        Overview and Motivation   64

        Assumptions and Terminology   65

        Conditions for Novelty   77

4   Applications   83

        Fresnel's Theory of Diffraction   83

        Special Relativity   86

        The Expansion of the Universe   93

        The Big Bang   94

5   Realism and Novelty   98

        The Burden of Argument   98

        Minimal Epistemic Realism   102

        The Explanatory Poverty of Instrumentalism   104

        In Defense of Abduction   116

        Novel Prediction versus Retroactive Explanation   120

        Partial Truth and Success   127

        The Pragmatics of Partial Truth   131

6   Counterarguments   136

        Overview   136

        The Skeptical Historical Induction   141

        Empirical Equivalence and Underdetermination   152

        Is There Truth in Virtue?   164

        NOA's Bark Is as Terse as It's Trite   173

7   The Future of Realism   178

        Limits to Testability in Fundamental Physics   178

        Methodological Practice   182

        Warranting Methodological Change   186

        A Measured Retreat   188

     Bibliography   191

     Index   195

Introduction

The more influential current theories of scientific epistemology are strikingly asymmetric in their assessment of scientific progress. The progressiveness of changes in empirical beliefs and of technological developments is regarded as relatively unproblematic. New forms of manipulation and control of natural processes improve over old; knowledge of macroscopic, experientially accessible portions of nature accumulates. By contrast, developments in theoretical explanations of observable phenomena are held to constitute not new or improved knowledge, but only more useful or comprehensive tools of inquiry. We are debarred in principle from knowing the true nature of entities, events, or processes that, at levels inaccessible to observation, operate to produce the facts we accumulate. Theories about the elementary constituents of matter or the overall structure of the universe are to be valued for their predictive success and formal elegance. But whatever the progress they achieve on such measures, theories do not extend our knowledge beyond the range of possible experience. They are not to be regarded as true or potentially endorsable accounts of the nature of reality; they are not representational.

It is only fair to locate this dichotomy within the legacy of logical positivism, to trace it even to the empiricism of the nineteenth century. For once theory and observation are divided, once experience is dissociated from interpretive thought, autonomy of epistemic stances becomes irresistible. Yet the current generation of scientific epistemologists does not, for the most part, have a positivist, philosophical agenda. Their skepticism or agnosticism with respect to theoretical knowledge is informed by, even grounded in, developments proper to science. Their arguments are not linguistic or conceptual, but epistemological. They think that knowledge can only go so far, and that how far it can go is something we know in much the same way that we know anything else; it is something learned from the scientific effort to advance knowledge.

In this book I try to develop a line of argument capable of contending with epistemological skepticism borne of science itself. I try to reinstate the common-sense idea that theoretical knowledge is achievable; indeed, that its achievement is part of the means to progress in empirical knowledge. My argument, too, is locatable within a legacy. It connects with William Whewell, and the proponents of the "method of hypothesis," in regarding explanation as a vehicle for the advancement of knowledge. But the conditions I regard as epistemically probative for theory are more specialized and arduous, in acknowledgment of the lessons of the recent history of knowledge. And the credence I recommend is as modest as it can be, once systematic skepticism of theory is disowned.

A NOVEL DEFENSE

of SCIENTIFIC REALISM

Truth and the Success of Science

Explaining Scientific Success

The academic study of science has reached adolescence. Having passed from wonder at the richness of our subject to the arrogance of disciplinary autonomy, we think that we know a great deal. But what we think we know keeps changing, and there is no consensus as to how we know it. We have many technical insights into details of the development of science; a few grand, if discredited, schemes for interpreting its methods and goals; and vigilant internal criticism. Still wanting is that consensus over fundamentals that emerges with the articulation of a tested methodology of inquiry.

Accordingly, almost anything of a general nature that I could say about science by way of beginning would be controversial. The ordinary image of science as the preeminently objective, disciplined, quantitatively rigorous means of building an ever broader and deeper structure of knowledge of the natural world has been challenged in every detail. In particular, and most remarkably, even the appraisal of science as a progressive and successful mode of inquiry is in jeopardy. Many are convinced that this complimentary picture is biased and self-serving, that the success of science is an artifact of the values and interests of a scientific culture.

It is possible to take this challenge seriously and to mount sophisticated arguments in defense of the objectivity and justifiability of the claim that science is successful and progressive.[1] My own response is deliberately simplistic. The thesis that such a claim is irremediably biased must, if correct, be biased itself. For as science is the paradigm of objectivity and intellectual progress, the thesis can only arise as an instance of imputing bias or interest-relativity to intellectual judgments in general, and as such, it must fall within its own purview—its own scope of application. But to the extent that a judgment is biased or self-interested, it is not to be trusted. Therefore, the imputation of bias to the judgment that science is successful and progressive is, if correct, not to be trusted. But of course, if incorrect, it is also not to be trusted. Therefore, it is not to be trusted. The thesis could be true, but it cannot be known to be true. There cannot be good reason to believe it.

It is confidently claimed by many historians and sociologists of knowledge that *all* research is biased. I wonder how this fact was discovered. It is not tautologous or otherwise self-evident. Evidently it took inquiry—that is, *research*—to discover it. But to the extent that research is biased, the conclusions to which it leads are untrustworthy. So this conclusion, that all research is biased, must, if correct, be untrustworthy. But of course, if it is incorrect, then it is also untrustworthy. Therefore, it is untrustworthy. It could be true, but we cannot have good reason to think so.

In a similar vein, it is popular to claim that knowledge is constructed. Sociologists studying the social forces at work within and among communities of research scientists have determined that scientific facts are not discovered but invented, and that the recognition of facts and their assimilation into bodies of knowledge are not learning but negotiation.[2] It is these social processes of construction and negotiation that determine what is taken to be factual, not interaction between investigators and the world. It is not the influence of independent objects of inquiry on the inquiring mind that fashions belief, but social forces. Given a fixed social structure, beliefs would emerge the same whatever the inputs from nature.

How did their examination of social practices enable historians and sociologists to reach these conclusions? If the conclusions are correct, then it could not have been the actual behavior of scientists, independently available for examination, that informed them. Rather, social scientists must themselves have invented or constructed these conclusions, and, subject to the social forces that dictate the methods and content of their research, negotiated the acceptance and dissemination of these conclusions among the ranks of social scientists. But if the conclusions were formed in this way, they do not at all convey information about what actually happens in science—about the content of, and standards for, scientific beliefs—any more than these conclusions allow scientific beliefs to register the workings of an independent, natural world. Rather, the claim that scientific facts

1. See, for example, Larry Laudan, *Science and Relativism* (Chicago: University of Chicago Press), 1990.

2. The seminal work is Bruno Latour and Steve Woolgar's *Laboratory Life* (Princeton: Princeton University Press), 1986.

are invented and their acceptance negotiated can register only the social norms operative within the community of social scientists.

My response to the claim that knowledge is constructed is to wonder how anyone knows this. Did he construct it? To say that something is constructed is to say that it is artifactual. It was made to be a certain way for certain purposes. Had the purposes been different, it would have been made differently. There was some choice as to the outcome. Well then, if the knowledge that all knowledge is constructed was constructed, it could be different. It could instead be the knowledge that some knowledge is not constructed (that being the only alternative). It could be true that all knowledge is constructed (or, at least, that what we *call* "knowledge" is constructed; there is a conceptual tension between the ideas of knowing and constructing, and the real intention is surely to deny that there is any genuine knowledge at all). But if it were true, it could not be known to be true. There cannot be good reason to believe it.

I used to think that there must be more than this to claims about the inevitability of bias, interest-relativity, and social determinism in intellectual judgment. I used to think that the facile dismissal of such claims that I have just given must be *overly* simplistic, that they require, or at least warrant, a more serious scholarly examination. But I have gotten over it.[3]

Accordingly, I shall dismiss such worries, and begin by assuming that modern natural science, at least, has succeeded in acquiring empirical knowledge. I do not allow myself assumptions as to the specific forms that scientific success takes or the scope of the empirical knowledge it achieves. Do we learn general truths about the world, or only effective means of manipulation and control? Does science reveal or potentially reveal everything in nature, or only some experientially accessible portion of nature? Such judgments must emerge from argument and analysis. I mean only to endorse the common and seemingly well-evidenced impression of science's cognitive productivity, vague though that impression may be.

What interests me is how this productivity is to be explained. A good deal of the product is itself explanatory; scientific theories and hypotheses are used to explain natural phenomena that, for reasons often rooted in previous science, have appeared puzzling or anomalous. But if we have now achieved, through science, a good understanding of the natural world, we have yet to achieve, through the study of science, a remotely comparable grasp of how understanding of the natural world is achieved. In contrast to science itself, which is as objective, disciplined, systematic, and well founded a mode of inquiry as ever there was, our understanding, or "science," of science in many areas remains at the level of homespun wisdom and anecdote. We only mask our ignorance by prefacing science texts with apparently authoritative chapters on scientific method, and by deploying a stock of ready clichés about what it is to be "scientific" in reaching or defending empirical beliefs. A little history of paradigmatically successful science quickly belies the most confident generalizations offered at this

---

3. Reading David Stove's *The Plato Cult* (Oxford: Basil Blackwell, 1991) helped.

level.[4] Just this fact, demonstrated so dramatically and frequently by historical studies of science, accounts for the rapid rise of interest in, and prestige of, the discipline of the history of science in recent years.[5]

Simplistic Accounts of Success

To give homespun wisdom and anecdote their due, some of what happens in successful science may be explained without sophisticated theorizing about science. Trial and error and the process of elimination no doubt suffice to account for some valuable results. Persistence and diligence in regard to detail are habits of mind or strengths of character associated with scientists that explain their success in applying these patterns of inquiry. Johannes Kepler discovered the elliptical orbits of the planets through a repeated failure to make compositions of circular motion fit the painstaking observational data for Mars compiled by Tycho Brahe. The explanation of this discovery lies in the exacting standards of Brahe's observations and in Kepler's systematic construction of various possible orbits from them. Nothing is mysterious about a success following upon a string of failures that issue from essentially the same methods or procedures. If the possibilities are limited, or circumscribable under some general description, persistence will eventually identify the right choice. If it frequently or usually happens that scientific procedures fail before they succeed, if for every advance there are many wrong turns or blind alleys, then perhaps there is little to explain. Perhaps the *impression* of the overall effectiveness and reliability of scientific methods results from the selectivity of our memories; success is what interests us and failure is quickly forgotten.

The most influential account of scientific method in our time portrays science in much this way. According to the *hypothetico-deductive method*,[6] hypotheses introduced to explain empirical data are tested against further data and eliminated if they

4. Among the popular assumptions about scientific method that most philosophers and historians have come to repudiate are (a) that individual hypotheses must be definitively falsifiable by observation; (b) that hypotheses are necessarily rejected if their predictions are false; (c) that hypotheses are inductively inferred from data; (d) that new theories are extensions of older theories to which they reduce "in the limit" of variables shared in common; (e) that scientific knowledge grows by accumulation, the conclusions of earlier science being retained by later science; (f) that observations and experimental results are impartial arbiters of theory; (g) that theories explain the successes of the theories they replace; and (h) that science is a purely descriptive mode of inquiry without normative or evaluative import.

5. Thomas Kuhn opened his influential book *The Structure of Scientific Revolutions* (Chicago: University of Chicago Press, 1962) by claiming that the study of science's history would revolutionize our understanding of its methods and goals. In this prophecy, if not in his substantive proposals to supplant that understanding, he is surely vindicated.

6. The influence of the hypothetico-deductive method within philosophy has diminished with the development of more sophisticated methodological theories. Within science, its influence on attitude, if not action, remains strong. The standard methodological pronouncements of science texts are squarely within the tradition of hypothetico-deductivism.

do not pass. Any number of hypotheses may be tried and eliminated before one passes. Even then, indefinitely much further testing is to be conducted, offering continuing opportunities for failure and elimination. It would seem likely that before any hypothesis has compiled a sufficiently good record to command confidence as an explanation, many will have failed. And then our success at finding an explanation, if or when it becomes appropriate to credit ourselves tentatively with that achievement, would seem a simple matter of creativity in the generation of prospective hypotheses and of diligence in testing them.

More sophisticated versions of the sort of procedure envisioned by the hypothetico-deductive method have been proposed under the heading of *evolutionary epistemology*. The idea is that hypotheses and theories face many stringent standards of evaluation and are forced into many forms of service. They are asked to produce new predictions; to fit consistently with already successful theories in diverse areas of research; to unify apparently disparate empirical phenomena; to help in the solution of problems other than those that motivate their introduction; to provide heuristic guidance in the construction of further theories; to help explain why theories they replace succeeded as well as they did. Such tasks become more onerous and more numerous as scientific knowledge increases. No wonder, then, that the relatively few theories to survive such rigors are fruitful and viable in application to scientific problems, and relatively stable through further research. Only the fittest survive. The explanation of the fact that science produces successful theories is just that it's a jungle out there, and theories that do not have what it takes to excel by the standards we use to reckon success, theories that cannot adapt to our ever-more-demanding needs and interests, die out rather quickly.[7]

But how do we manage to invent any theories at all that present the virtues we demand? How is even *occasional* success to be explained? To the extent that success is occasional only, "luck" or "chance" might seem a credible answer. After all, if, for all our effort, we were *never* successful, *this* would be the mystery. But the evolutionary epistemologist has an allegedly deeper analysis: We can deduce that we are successful just from the fact that we are here. Beings like ourselves, beings reliant on intellect, who failed even to stumble often enough on theories that work, at least at the level of prediction and control of nature, would die out. No doubt many have, for just this reason. But we, as a culture with established, ongoing scientific practices, survive. Therefore we do, whether by stumbling on them, or in some more reliable way, come up with theories that work. We have evolved certain processes of theory construction and selection. This means that lots of processes are tried, but only effective ones are retained because beings using ineffective ones do not survive to continue them. Our processes must be effective, if they are still in use.

7. This is Bas van Fraassen's position in *The Scientific Image* (Oxford: Oxford University Press, 1980): "I claim that the success of current science is no miracle. It is not even surprising to the scientific (Darwinist) mind. For any scientific theory is borne into a life of fierce competition, a jungle red in tooth and claw. Only the successful theories survive" (p. 40).

Now, to deduce is, with caveats, to explain. There is a long tradition assimilating explanation to the deduction of what is to be explained from what is better or already known. If we can show that something is inferable from what we know, then we can show that it is to be expected; it should occasion no surprise. At least it presents no further explanatory task beyond that posed by the knowledge from which it is inferred. Thus it is concluded that our very existence explains our success. The reason that we have successful science is that all who do not have it are dead; that is, they are other than us.

This line of thought may appear convoluted. If scientific success makes the difference between survival and extinction, then the fact of survival cannot explain success. Perhaps deduction can go in both directions between success and survival, but explanation, on pain of circularity, cannot. This is a qualm about evolutionary explanation in general; let us not trouble over it. For even if an evolutionary response to the challenge to explain success is correct, it cannot meet this challenge fully. It no more stands on its own than does evolutionary theory without molecular biology.

Evolution deals statistically with populations; it is silent as to the individual case. It explains not why particular birds fly south, but why the attribute of flying south tends to be promoted over time among populations of birds. To the question asked about a bird that flies south *why* it does, evolution gives an indeterministic answer; that is, its answer is true also of birds that do not fly south. It only locates the bird in a population within which flying south is a beneficial trait, and that population may include birds that fly north. At the level of the individual we need a structural analysis capable of distinguishing one individual from another, such as molecular biology (potentially) supplies.

Similarly, if we ask of successful theories *why* they are successful, we need an answer that goes beyond an explanation of why science in general produces successful theories; we need an answer that appeals to attributes that discriminate among theories. Why does *this* theory work, while others equally the products of diligence and preferred methods fail?

Notice that it makes no essential difference to appeal to forms of discrimination that simply increase the stringency of the hurdles a theory must pass. We have learned to identify certain characteristic errors that admit bias or conceal relevant alternatives in the testing of theories. We have learned to control experiments so that a hypothesis will not be confirmed by results that could equally well be taken to confirm a rival hypothesis. We conduct blind experiments that eliminate effects capable of biasing the results. Larry Laudan argues that such measures suffice to explain why science succeeds.[8] But it does not explain why a theory correctly predicts the results of a controlled, blind experiment to be told that the experiment had these checks built in. These checks may explain how we come to identify useful and reliable theories, but they do not explain why the

---

8. Larry Laudan, "Explaining the Success of Science: Beyond Epistemic Realism and Relativism," in *Science and Reality*, James Cushing, C. F. Delaney, and Gary Gutting, eds., (Notre Dame, Ind.: University of Notre Dame Press, 1984).

theories we identify as useful and reliable have these virtues. It is, at most, the success of science *as a human activity* that evolutionary hurdles explain; they explain why *we* develop successful theories. But why are successful theories developed at all? The success of science *as a body of theories*, the success of the particular ideas that survive those hurdles, is unexplained.

Laudan trades on an ambiguity that he himself has plainly noted.[9] Suppose I am asked why certain individuals, say those observed during a telecast of the Wimbledon finals, are such great tennis players. There are two ways to take the question, distinguished by whether the identification of the individuals in question as Wimbledon finalists is incidental or essential. To explain why *Wimbledon finalists* are so great, it is perfectly appropriate to cite the stringency of the selection procedures for entry into the tournament; the fact that the prestige, prize money, and ancillary financial opportunities gained by victory are such as to draw the best players; and the rigors of the tournament schedule by which all but two players have been eliminated. It is hardly surprising that the finalists are great players, considering what they had to go through to get there. However, none of this explains why *these particular individuals*, who happen to be the finalists, Bjorn Borg and John McEnroe, say, are so great. On the contrary, it is their being great that explains their having managed to survive the rigors of selection. To explain why Borg and McEnroe are so great, we must cite *their own* relevant properties, their training and genetic attributes, for example, not the standards they face in common with the less successful.

Analogously, to explain why the theories *that we select* are successful, it is appropriate to cite the stringency of our criteria for selection. But to explain why *particular theories*, those we happen to select, are successful, we must cite properties *of them* that have enabled them to satisfy our criteria. Laudan argues as though the fact that our explanatory question can be given a reading on which stringency of criteria for selection is the right answer obviates the need to consider attributes of successful theories themselves. He surreptitiously dismisses the form of the question in which the identification of the theories whose reliability we are to explain as those that have met our standards is incidental.

Dismissive Attitudes toward Explanation

Perhaps this further explanatory question has no answer. Perhaps success at the level of the individual is a random event. Instead of evolution, our model might be quantum indeterminacy. Then *nothing* explains why *this* theory works, any more than anything explains the choice of the helium nuclei to be ejected in radioactive decay. It is just an irreducible fact of nature that certain ways of representing the structure of the world predict correctly while others do not.

Closely associated with this view is a dismissive attitude toward explanation as

---

9. See Laudan's reply to criticism in the symposium "Normative Versions of Naturalized Epistemology," *Philosophy of Science*, 57 (1990): 44–60.

a vehicle for advancing knowledge. One justification offered for this attitude is that the truth of an explanation is incidental to its status as an explanation. One does nothing to improve the explanatory power of a theory proposed to explain something by adding that the theory is true. If it is the kinetic energy of gas molecules that is to explain the pressure a gas exerts on its container, nothing explanatory is gained by supposing that the molecules are real; their reality is not part of the explanation. According to an example of Ian Hacking's, Albert Einstein's use of photons to explain the photoelectric effect does not involve imputing existence to photons, nor does the fact of their existence, once learned, contribute to the explanation Einstein has already achieved.[10] In a similar vein, Michael Levin argues that explanation proceeds by providing a mechanism for the production of the effect that is to be explained, and truth is no mechanism. "By being true" never answers the question how a theory manages to explain or predict anything, says Levin.[11]

Many philosophers find this line of reasoning convincing, but I have great difficulty appreciating it. Let me try to appreciate it via an analogy. I do not show that someone should be convicted of a crime by showing that he was able to commit the crime. The fact that he was at the scene or could have been, and had the skill or equipment necessary to commit the crime, is compatible with innocence. Analogously, the fact that the mechanism a theory describes is capable of producing the effect I want to explain does not imply that that is the mechanism responsible. So I cannot deduce the theory's truth from its explanatory power. If, therefore, I claim truth for the theory, my claim is strictly independent of an assessment of the theory as an explanation.

I do, however, show that the defendant is innocent by showing him incapable of committing the crime. Of course, he could still be guilty of some other crime. Similarly, I show a theory false *as an account of the particular effect to be explained* by showing the mechanism it posits to be incapable of explaining this effect, although the theory could still be true and correctly account for other effects. So it is at least *part* of a case for the truth of a theory based on inference from a certain effect that the theory *not* be incapable of accounting for the effect. Establishing the defendant's ability is a necessary, evidentially relevant, part of a case for conviction. If I show further that the defendant is *uniquely* capable of having committed the crime, then I establish guilt. And to the extent that I discredit alternative mechanisms to the one my theory posits, or discredit the possibility that there are alternative mechanisms capable of producing the effect, I justify imputing truth to my theory. Should we not then allow, at least, that a theory's explanatory power is evidentially relevant to the justifiability of imputing truth to it? Whether or not truth is a mechanism, the capabilities of the mechanism—its explanatory power—are vital to the assessment of truth. If the point

10. Ian Hacking, *Representing and Intervening* (Cambridge: Cambridge University Press, 1983), p. 54.

11. Michael Levin, "What Kind of Explanation Is Truth," in J. Leplin, ed., *Scientific Realism* (Berkeley: University of California Press, 1984).

Hacking and Levin are making is simply that evidential relevance is not evidential sufficiency, they have overstated it. They have merely called attention to the problem of *underdetermination*, the problem (roughly) of discrediting alternatives. They have not succeeded in divorcing truth from explanation. What further point they might be making, or might be made for them, I shall consider in the fifth section of this chapter.

Another reason, championed, for example, by Hacking and Bas van Fraassen, for a dismissive attitude toward explanation is that explanation has the practical function of identifying features of a situation that bear on our interests. What makes an explanation right is that it gives us what we need to solve a problem in some context. No explanation is uniquely right. Thus the right explanation of why a building is on fire is different in the context of fixing criminal liability from what it is in the context of ensuring public safety. More generally, an explanation tells us something about *ourselves*, about our needs and interests; it does not, properly understood, increase our knowledge of an independent natural world.

Evolutionary processes explain by connecting patterns that interest us in natural history with traits we can observe or measure. But the potential circularity of such connections shows that evolutionary explanation presupposes one among alternative options for singling out an explanatory task. The facts as such are not inherently explanatory. Explanations do not belong in our ontology, along with the events and processes that they relate. Once discovered, facts can be put to explanatory use, depending on what interests or purposes of ours they relate to. But the search for explanations, the posing of explanatory tasks, does not lead to knowledge; that is to get the process backward. There are no context-free, interest-independent, right explanations to be found, including explanations of why some theories work and others do not.[12]

Admittedly, 'explain' is a highly prudential, interest-relative term in its ordinary use. But I do not think that reflecting on what would satisfy as answers to such questions as why a building is on fire provides the proper model. There is a sense of 'explain' in which something can be explained even if no surprise is diminished, no comprehension imparted, no practical interest served. It was once popular to say that only three people understood the general theory of relativity. Be that as it may, there was surely a time (in 1915) when only one person understood it, and there could have been a time, after its invention, when no one did. (Einstein could have lost his grip; he once said that he had thought he understood special relativity until Hermann Minkowski reformulated it.) Surviving such possibilities is a straightforward sense in which general relativity explains gravity. This sense does not seem to be contextual or interest-relative.

The dismissal of the question of why a theory succeeds seems to me most unsatisfactory. Does it seem so because *I* have some need for, or interest in, explaining why a theory works, for want of which need or interest there would *be* no explaining why it works? But this suggestion can be turned on its head. In wanting to know *why* it is

12. Thus, Hacking dismisses explanatory success as "what makes our minds feel good." See *Representing and Intervening*, p. 272.

theory T_1 *rather than* theory T_2 that works, I am indifferent as to whether it *be T_1* rather than T_2 that works. Of *whichever* theory it happens to be, I wish to know why *that* one. An answer relativized to my interests could not discriminate between the theories, as an adequate answer must. The view in question therefore reduces to the position that there is no explanation at all.

This hardly seems plausible as an initial position. Perhaps we will come to it out of desperation, but what is its motivation *ab initio?* The evolutionary model does not ground it. Evolution was never supposed to be the whole story, and has become extendable by microbiological explanation of the individual case. The quantum model does not ground it, because its injunctions as to the individual case are disanalogous. There is no explaining *or predicting* individual quantum events. But we surely can predict what theories will work. They are the ones that have worked thus far.

This is a feature of scientific experience missing from hypothetico-deduction. That methodology treats each new confrontation of theory with test-result like the last. A theory can fail at any time, no matter how many tests it has passed. There is no provision for treating a seasoned theory, with many passes to its credit, differently than a fledgling theory; failure is always decisive. But we trust theories better the more tests they have passed, and we have found it generally reliable to do so. Not only is our expectation of success greater for theories successful already than for new ones, but we are less inclined to regard the failure of a successful theory as sufficient grounds for rejecting it. We will seek other ways to explain the failure and to protect the theory from it. This effort often works. That is why general methodologies for science opt for a more robust form of inductive support for theories, a more robust form of confirmation, than the simple accruing of positive test results that hypothetico-*deduction* allows. As we can reliably predict what theories will fare better in confrontation with new tests, and what theories will prove salvageable when a test is failed, it is reasonable to call for some explanation of their relative success.

Underdetermination

Of course, it can be reasonable to call for an explanation without its being possible to provide one. Rejecting, or deferring as the evil of a later day, the view that there is no explanation, we can consider the possibility that the explanation that there is is one we cannot discover. This position is based on the problem of *underdetermination*, a problem rooted in the limitations of the hypothetico-deductive method. Hypothetico-deduction is first and foremost a method of *disconfirmation*. As a method of confirmation, it says nothing more than that so many tests have been passed. It tells us when to reject a theory, but not when to accept one. Implicit in this priority is a recognition that tests can be potentially disconfirming for theories without being potentially confirming. Passing a test is not in itself confirmatory if the test is one that even a false theory would be expected to pass. If the prediction by which a theory is tested is one whose correctness we can explain independently, then the fact that the prediction is borne out lends no special support to the theory. Failure of the prediction still discredits the theory, however, as it does whatever other basis there is for the prediction, because prediction is a matter of *entailment:* The truth of the theory, along with presumably well-established further assumptions, entails the truth of the prediction.

The thesis of underdetermination is a kind of massive generalization of this situation. It denies that theories can be individuated solely by means of their testable consequences. No amount of empirical evidence, according to this thesis, singles out any theory as uniquely successful. Instead, there will always be a plurality of theories that conform to the test results. Perhaps some are generated by algorithmic operations on others; perhaps some are unformulated, not yet entertained. The thesis need not be read as a historical description of scientific practice; its charitable reading speaks of "possible theories." It insists, in any case, that while we can disqualify many theories by hypothetico-deduction, we cannot disqualify enough. Such a process of elimination will always leave too many theories in the running for us to be able to identify the correct explanation of the test results.

If the thesis of underdetermination were correct, it would seem that *no* potentially disconfirming result is potentially confirming; passing a hypothetico-deductive test *never* carries confirmatory significance, whether or not the test is one that we have reason to expect even an unacceptable theory to be able to pass. Or, to put it another way, the sort of situation we imagined as defeating the confirmatory significance of passing would appear, under conditions of underdetermination, *inevitable.* As more than one theory correctly predicts the result of any test, there is always a basis for predicting the result independently of any particular theory being tested. The significance of passing, then, is at best to increase the likelihood that the theory is *among* those that accord with the observable phenomena. At best, it belongs to a group all but one of which are wrong and no one of which, therefore, is likely to be right. The maximum probability that any theory can have is ½, as the minimum number of theories compatible with the evidence is two. There is an explanation of what we observe, but as its probability cannot be greater than that of competing explanations, it cannot be discovered.

The cogency of the inference from underdetermination to the impossibility of confirmation, and, for that matter, the cogency of the underdetermination thesis itself, are issues we may defer.[13] Our immediate question is whether, in addition to allowing us an unknowable explanation of what is observed, this view allows any explanation, knowable or not, of our *success at predicting* what is observed. The distinction here parallels the one I used to show the incompleteness of Laudan's explanation of success. It is subtle, and care must be taken to characterize the explanatory task in a way that discriminates between its terms. On the one hand, there are the phenomena predicted by the theory, the obtaining of which is a success for the theory. These phenomena we wish to be observable (or other-

---

13. See chapter 6, under "Empirical Equivalence and Underdetermination."

wise unproblematically attestable), so that we are in a position to judge that they do obtain and so to pronounce the theory successful. On the other hand, there is *the theory's success in predicting the phenomena, the fact that this particular theory manages to predict successfully.*

In asking for an explanation of a theory's successful predictions, we may be asking only for an explanation of why *these particular phenomena*, the ones the theory happens to predict, obtain, rather than something else that could have happened. That is, the description "the theory's successful predictions" may be used to identify the phenomena at issue but be *incidental* to these phenomena; it is these phenomena we want explained whatever theory predicts them, perhaps whether or not *any* theory predicts them. But in asking for an explanation of a theory's success in predicting the phenomena, or (more conveniently) of a theory's predictive success, the identification of the particular theory mentioned occurs essentially. We want to learn, not of these predictions why they are correct, but of this theory why *it* predicts correctly or why *its* predictions are correct. It might of course be objected that this latter question is somehow illegitimate, trivial, or without answer, but it is in any case a *different* question. It is the question I mean to address in pressing underdetermination as to the explanatory resources it allows.

Does a theory's membership in a group of predictively indistinguishable, successful theories—which, according to underdetermination, is its only potentially confirmable status—do anything to explain its success thus construed? What is the explanatory status of such membership? Presumably, one member of the group *does* explain the success of its predictions. This is the true one. (Perhaps it is only a potential member; recall the charitable reading of underdetermination.) This theory itself, that is, its own theoretical pronouncements, explains why certain phenomena predicted by it obtain. And its truth, or the truth of these pronouncements, the fact that the structures or processes it identifies are actually operative in producing those phenomena, explains how it manages to predict successfully. Other theories in the group predict successfully because what they predict happens to agree with what that one predicts.

But is it any explanation of a theory's predictive success that its predictions *happen* to be the same as predictions that are explained but whose explanation is unknown? This tells us that a theory is successful because it predicts what the true theory predicts. But why does it do that, unless it itself is the true theory? It seems that only in that one case, which underdetermination claims to be unidentifiable, do we have an explanation. Unless the members of the group share more in common than their observable predictions, there is nothing about them to cite by way of explaining why they share *that* in common, why those that are not true give the very predictions given by that which is true. If the false ones just *happen* to agree with the true one, there is no explanation of their success at all. What *just* happens is necessarily unexplained; explanation requires a cause or reason for happening.

The alternative, then, is that the false, successful theories bear some connection or similarity to the true one beyond predictive success, which explains their predictive success. Perhaps they posit similar structures or processes, and are thereby *approximately* true, or *partially* true, or at least *not completely false*, to give the similarity a suggestive if tentative description. But whatever the similarity turns out to be, it should be clear that there must be one or there simply is no explaining predictive success under circumstances of underdetermination (and its purported implications). There is an explanation of why the *predictions* are successful, which is unknowable, but there is no explanation for any theory but the true one of why *its* predictions are successful, of how *it* manages to entail the right predictions. To the extent that underdetermination is found to be a genuine problem, only the imputation of truthlikeness, in some form, to successful theories avoids giving up altogether on the explanatory challenge I have identified.

The Alleged Superfluousness of Truth in Explanation

I am arguing that a theory's success is explained only by somehow relating the theory to a true theory, that explanations of the success of false theories are parasitic on the true theory. In so arguing, I have supposed that the truth of a theory is, *in and of itself*, explanatory of its success; if, somehow, we were able to establish that a particular theory is true, there would be no further explanation wanting of why it is empirically successful. This is not to assume that the truth of a theory *entails* that it will be empirically successful. Diagnosing a particular disease explains symptoms characteristic of the disease even if it is possible to have the disease without displaying those symptoms. The truth of a statistical theory cannot guarantee that the results of testing it will conform to the expectations it induces. And any theory could be unsuccessful even if true, through difficulties or errors in its application. Nor do I assume that any theory, true or not, need *be* testable; perhaps some theories just do not yield observable predictions. But when a theory *is* empirically successful, I have supposed that there is an explanatory task that its truth would, if determinable, fulfill.

I have acknowledged that some philosophers hold exactly the opposite view, contending that a theory's truth adds nothing to whatever explanation, if any, may be given of its success, that truth is superfluous in explanation. Let me explore some possible reasons for holding this view, without regard to whether they are operative.

Most obvious is the question of burden of proof. I described my own invocation of the explanatory force of truth as a *supposition*. Perhaps I am the one who owes reasons, rather than those who do not make that supposition or who deny it. It seems to me, however, that we are entitled to assume, in a context in which the success of a theory is at issue, that we do have in the theory itself at least a purported or potential explanation of phenomena by which it is tested. A theory is not simply an empirical law or generalization to the effect that certain observable phenomena occur, but an explanation of their occurrence that provides some mechanism to produce them, or some deeper principles to which their production is reducible. Theories are inherently explanatory in that they propose answers to questions or solutions to problems that antecedent science raises. Generally, it is the correctness of a theory that is at issue in theory appraisal, not its capacity if true to explain the phenomena it predicts.

Of course, a theory proposed to account for some type of phenomenon might turn out to be acceptable, if at all, only as a theory of some subportion of that type. For example, Niels Bohr's original theory of the hydrogen atom proved unable to account for many properties of hydrogen spectra that an adequate theory should have explained. James Clerk Maxwell's electromagnetic theory of light could not account for all the known wavelike properties of light, let alone its particlelike properties. In such cases a theory may be criticized as inadequate to explanatory tasks, without regard to its truth. But then the phenomena that comprise explanatory tasks at which the theory has failed are not among those by which it is tested. So we are still entitled to assume that a theory proposes some explanation of the phenomena it predicts and by which it is tested. The question of whether it *does* explain those phenomena, whether it actually explains them or merely potentially or purportedly explains them, is primarily a question of its truth.

Therefore, no question of accounting for the success of a theory presumed true arises. The theory manages to predict correctly because it has identified the mechanism or principles actually operative in producing the phenomena we observe. The question is why a *false* theory should predict correctly. If the mechanisms or principles a false theory posits are inoperative and unlike those that do operate, why should predictions from them be correct?

The question of how the wrong theoretical structure yields the right predictions can invite an answer that maintains the superfluousness of truth in explanation: It does so by *deduction*. It might be argued that the deductive structure of prediction explains a theory's success, that the challenge to account for the success of false theories, without invoking truth through similarity or approximation, is met simply by noting that from the false theories the correct predictions are deducible. A theory manages to predict successfully by providing a basis for the derivation of descriptions of successfully predicted phenomena. This answer trivializes the question of why a theory is successful, but that may be taken as a criticism of the question rather than as a defect in the answer.

It is important to be clear about the nature of this trivialization. It is not a point of logic. Sometimes it is said that from false statements *everything* is deducible, as if the falsehood of a theory, by itself, could explain the theory's ability to yield wanted predictions. I have heard noted philosophers claim that the fact that a theory is false suffices to explain its predictive success, as though empirical success is as well, or better, taken to indicate falsehood as truth. This is simply a mistake. It is not from falsehood, but from *necessary* falsehood, or inconsistency, that everything is deducible. The fact that a statement is false has no bearing whatever on the statement's implicative capacities.

The triviality in question results, rather, from the fact that prediction just *is* deduction (with the usual caveats for statistical inference); one makes predictions on the basis of a theory by determining the theory's observational consequences via "auxiliary hypotheses"—that is, background theories, collateral empirical information, and other assumptions allowable in context. With proper specification of such background knowledge—empirical determination of initial conditions used in applying theoretical laws, for example—prediction has a deductive structure, although what is deduced may be an empirical regularity that holds only statistically. Therefore the answer "by deduction" works as well for the true theory as for the false ones, thus implying that truth is incidental to the explanation. The suggestion is that in showing, empirically, that a theory's predictions are successful, and in showing, formally, that they are *its* predictions, that they bear an appropriate derivational connection to the theory, we explain all that needs to be explained about any theory's success. A theory manages to predict by implying; it manages to predict successfully by implying what is found to be the case.

The trouble is that it is then just a *coincidence* that different theories, positing diverse mechanisms for the production of the predicted phenomena, all manage to do that implicative work, and to do it successfully. And while it admittedly could just be that this is a coincidence, a philosophical position that requires such an explanatory lacuna does not, once again, seem reasonable as an initial position. It would perhaps become reasonable when accompanied by some reason for denying that an explanation exists, if only a reason founded in desperation, but it is not reasonable *now*.

A deeper objection to the trivial answer is that it completely separates the question of how a theory predicts from the question of the truth of what it predicts. It treats the correctness of the predictions as incidental. But the question to be answered in explaining a theory's success is not how the theory manages to predict certain phenomena, phenomena that happen, parenthetically, to be observed; the answer to this question *is*, no doubt, just "by deduction." Rather, the question is how it happens that it is the *observed* phenomena that are predicted, where 'observed' enters essentially into the identification of the phenomena in question. "By deduction" does not answer this question, for nothing in the process of deduction as such carries any implication for the correctness of what is deduced.

Another possible basis for a dismissive attitude toward truth in explanation is found in the *redundancy theory of truth*. The basic idea here is that predicating truth of a proposition amounts to no more than *asserting* the proposition. Attributions of truth are superfluous because the conditions for the correctness of the attributions are given just by the propositions to which the attributions are made. The suggestion of a thoroughgoing redundancy theory is that the very concept of truth can be analyzed away; it is unnecessary because anything said by using it can be said as well without it.

In this form the redundancy analysis is certainly untenable. A small indication of its difficulties is already evident in the formulation I have given. One must use the concept of truth (or correctness) in showing that attributions of the concept are unnecessary. In general, second- and higher-order uses of the concept, in which the term 'true' (or its cognates) appears iterated, will defeat the analysis. So will attributions of truth to indefinite disjunctions, such as those that the underdetermination thesis envisions; we might need to assert that some theory (or other) is true without being in a position to assert any theory or disjunction of theories.

Such objections do not defeat the point of invoking the redundancy analysis, however, and the analysis may be applied in a restricted form. It may be said that theories do their own explaining; nothing is added to the explanations a theory itself provides by attributing truth to it, because such attributions are collapsible into the theory's own assertive force. Theories make ontological commitments; they tell us what kinds of things there are, what they are like, and how they work. They explain by positing mechanisms for the production of what is to be explained. Attributions of truth do not help to explain, because, once again, truth is not a mechanism.

One objection I have to this line of argument is that the ontological commitments of theories cannot always be read definitively off the theories themselves. Just what status a theoretical construct or posited entity is to have often is a matter of philosophical interpretation. Such inherent ambiguity is by no means the creation of quantum mechanics, the most famous case of a theory subject to competing ontological interpretations. The question of when a newly introduced term that is given a grammatically referential role in a theory is to be taken as purportedly referential, or genuinely referential *according to the theory*, arises throughout nineteenth-century physics. It arises for the concept of potential energy, for the concept of the electric field, and for the concept of the mechanical ether, to take the more obvious examples. It arises for the concept of the space-time interval in relativity theory. Nothing in the theory itself tells us whether this is just an abstract mathematical quantity, or is a replacement for space and time, which are henceforth to be regarded as the abstractions. Attitudes on such matters are overlays on the theory and can change radically while the theory remains intact. The adage "Maxwell's theory is Maxwell's equations," taken to mark the abandonment of attempts to interpret the ether mechanically, really does more to *raise* the problem of ontological ambiguity than to solve it; it attempts to expel the very issue of what the equations are talking about from the theory proper.

I will put aside the issue of the extent to which theories speak for themselves, for there is a more direct and definitive objection to the use of a redundancy analysis to obviate the role of truth in explanation. What the analysis obviates, if anything, is not truth, but 'true'. Even if it shows that we need not *speak* of truth or use the term 'true', or its cognates, in explaining, it does not show that we can succeed in explaining without speaking truly. The point I have been urging is that only for true theories, or theories like them, do we have an explanation of predictive success, not that such theories must be *called* "true" or "partially true" to give that explanation. It is not the concept of truth, but truth itself, for which I claim a role in explanation. Truth is crucial to science, even if the concept of truth need have little role in understanding science.

But in fact, the redundancy analysis, taken on its own terms, does not eliminate even the concept of truth from explanation, for it identifies this concept with something retained in explanation, viz., the theory's own assertions. Note that it is not truth simpliciter that the analysis reduces to assertion, for assertions may be false. It is, at most, *the assertion of an assertion's truth* that is reduced to the assertion itself. And I grant that we need not assert a theory's truth to explain its success; we need only assert the theory. But this we may not do unless the theory is true. 'True' may be superfluous, but not truth.

Another possible reason, or motivation, for dismissing truth from explanation is despair of discovering what the true theory is. Perhaps underdetermination, inadequacy of evidence, or some other incapacity in the methods of theory construction and evaluation forever impedes our quest for the truth. A requirement for truth in explanation comes to nothing if it can never be met.

As it stands, this point looks much like an instance of a familiar, if infrequently identified, fallacy, the fallacy of confusing an *epistemological* point with a *metaphysical* one. It is an epistemological point, a point about the kind of knowledge available to us and the conditions for obtaining it, to deny the possibility of learning what theory is true. The requirement that explanations be true, by contrast, is a metaphysical point about the nature of explanation itself, regardless of whether or how it may be achieved. If explanation requires something unavailable, then explanation is unavailable. If knowing that a purported explanation is true is a requirement for knowing that it is a genuine explanation, and if we cannot know what is true, then although we may *have* a genuine explanation we cannot know that it is genuine. It does not defeat the status of a condition as a requirement to show that it cannot be met. In practical affairs, it may be shown that it is unwise to *institute* a requirement by showing that it cannot be met. But we are concerned here not with instituting a requirement, but with *discovering* what is already, in the nature of the case, required.

The point may be reformulated to avoid this problem. If we assume that we do have explanations and know that they are explanations, we are then entitled to infer that no unmeetable condition can be required for explanation. In particular, if the truth cannot be known, then it cannot be required that we know anything to be true in order to know that it is explanatory. It still would not follow that explanations do not have to be true, only that they do not have to be known to be true. In general, a necessary connection between two things, whereby one is required for the other, is compatible with independence of the knowledge of those things.

For example, there may be a necessary connection between water and oxygen, in that nothing could be water without containing oxygen. Yet, it can be known that something is water without its being known to contain oxygen. After all, people reliably identified water before the discovery of oxygen. Descartes thought that he could exist without physical properties, that physical properties could not be essential to his existence, because he knew that he existed without yet knowing whether he had physical properties. This is a philosophical mistake; an adequate basis for knowledge of the existence of something need not include information as to all the properties without which the thing could not exist.

If, however, we know that something is explanatory and know that nothing could be explanatory without being true, we could infer that it is true, adding this point to our knowledge. So if knowledge of truth is really impossible, knowledge of explanatory status becomes impossible as well if it is known to require truth. Now suppose that if explanation requires truth, this fact can be known—a supposition that it is only reasonable for me to allow as I have lately been engaged in establishing the fact. The supposition that we do know explanations then blocks the requirement of truth.

Faced with the alternatives this reasoning poses, we may well wonder whether its conclusion represents the best choice among them. Is it not as reasonable to question the assumption of knowledge of explanations as it is to reject truth as a requirement? Perhaps what we think we know to be explanations are really unsuccessful, purported explanations. Is it not as reasonable to allow the possibility of learning the truth as it is to block knowledge of explanations, or truth as a requirement for them? The answers depend on the reasons given to trust our presumed knowledge of explanations or to deny the possibility of learning the truth. I have provided reasons for insisting that truth is required, but we have not yet seen any comparable grounds for denying that the truth can be known. There is only the thesis of underdetermination and its alleged consequences, which we have deferred.

Consider an analogy. One often hears it argued that there is no such thing as morally right or wrong action, because of the difficulty in *proving* that any action is morally right or wrong. (This line seems particularly popular with students and political scientists.) Implicit is the assumption that nothing can be right or wrong (in "absolute" terms—subjective, or otherwise relativized, right and wrong are generally allowed but are meager compensation) without it being knowable that they are, which is an instance of the fallacy of confusing the metaphysical with the epistemological. It is further assumed that knowledge requires proof (in the robust sense of argumentative cogency that originates in mathematics). This assumption is completely untenable. If proof were required to know anything, there is precious little that many of us would know; not much of *common* knowledge would be knowledge. The mathematical proof that certain propositions can be known only if they cannot be proved preempts any identification of the knowable with the provable. In any event, it is *at least* as evident that there is a (nonsubjective) difference between right and wrong as it is difficult to provide proofs. I suggest, analogously, that explanation requires truth whether or not truth is knowable, and that it is at least as evident that explanation requires truth as it is difficult to establish what is true.

Reasoning similar to that which I have been criticizing has been used to oppose the classic "deductive nomological" model of scientific explanation proposed by Carl Hempel and Paul Oppenheim.[14] On this model, explanations have the form of deductive arguments with nomological premises. This model, in effect, presumes that explanations must be true, in that it requires the laws used in explanations to be true, or at least highly confirmed. Reasoning is not explanatory unless laws cited as premises meet this condition. To this requirement Laudan has objected that the way we decide whether a proposed law, or, more generally,

---

14. Carl G. Hempel and Paul Oppenheim, "Studies in the Logic of Explanation," *Philosophy of Science*, 15 (1948):135–175; reprinted in Carl G. Hempel, *Aspects of Scientific Explanation* (New York: Macmillan, Free Press, 1965).

a hypothesis, is true, the way we establish that it is well confirmed, is by evaluating its success at providing explanations. We must therefore be able to tell whether a hypothesis is successful at providing explanations independently of knowledge as to its truth. Truth cannot be required for explanation because we must be able to know that a hypothesis is explanatory without (yet) knowing that it is true.[15] The objection does not work, because, as I have tried to show, there may be a connection between truth and explanation that does not hold between knowledge of truth and knowledge of explanation.

A more subtle form of the objection might yet succeed. If hypotheses had to be true to be explanatory, then we could not know this without being in a position to know also that hypotheses used in explanations are true, thus obviating the methodology the objection supplies for evaluating them. That is, if we knew that explanations must be true, then we could simply infer from the presumed explanatory status of our reasoning that any hypothesis functioning as an essential premise of it is true. As truth cannot be established in this way, we cannot know that explanation requires it.

But we can instead conclude that our presumptions of explanatory status are wrong. A more reasonable way to account for evaluative practice than by dismissing truth from explanation, it seems to me, is to say that we evaluate the relative effectiveness of purported explanations as a means to judging which hypotheses to accept. There are many dimensions of appraisal for purported explanations that do not involve deciding whether they *really* explain, and this is all the independence of explanation from truth that the methodological point of the objection requires. We assess comprehensiveness, the quantitative accuracy of prediction, simplicity, and heuristics.[16] And we compare rejected or superseded attempts at explanation as to superiority; evidently, such comparisons are independent of judgments of truth. I think, for example, that we should say that Issac Newton's *way of explaining,* or his *proposal to explain,* the tides was better than Galileo's, although from a relativistic perspective, neither really explained them. Thus the evaluations we make need not presuppose explanatory status, as the objection assumes.

Excising Truth from Explanation

There is yet another possible reason for denying that the explanation of theoretical success requires imputing truth (in some measure) to successful theories. It may be thought that a weaker assumption can do all the explanatory work of truth while avoiding its metaphysical commitments, which exceed our epistemic capacities. This approach exists in two versions; one may be associated with van

15. This reasoning underlies Laudan's dissociation of truth from explanation, or problem solving, in *Progress and Its Problems* (Berkeley: University of California Press, 1977); see p. 22.

16. For another argument, see Jarrett Leplin, "The Assessment of Auxiliary Hypotheses," *British Journal for the Philosophy of Science,* 33 (1982):235–249, esp. 246–249.

Fraassen, and one is due to Arthur Fine. Both philosophers deny that the need to assert a statement in giving an explanation is good reason to believe the statement; neither would allow inference to the truth of a statement from its explanatory status. But further, each holds that theoretical statements *need not be asserted* in explanations of empirical phenomena, or in explanations of their own success at predicting empirical phenomena; any explanation achieved by means of theoretical claims may be replaced by equally good explanations that are noncommittal as to the truth of those claims. To construct the replacements, a general rule or operator is provided that excises theoretical commitments.

The result is a kind of double attack on the view I am defending: First, we do not need to attribute truth to theories to explain anything; second, even if we did need to attribute truth to theories to explain, this would not justify believing that they are true. These claims may not be clearly distinguished, although they are in fact independent. It is the first claim that concerns us immediately as a threat to the connection of truth with explanation on which I have been insisting. The second claim will occupy us later.

What van Fraassen proposes to replace a theoretical claim is a generalization of the claim's observable consequences.[17] Of course, observable consequences are deduced from complexes of theoretical claims, in a rich context of auxiliary information. So it is more plausible to speak of the consequences of a theory, as an open-ended structure of propositions, although the question of truth applies directly to statements individually. If the observational consequences of a theory are all true, van Fraassen calls the theory "empirically adequate." He contends that any explanatory or predictive achievements of a theory are as well explained by asserting that the theory is empirically adequate as by asserting the theory itself. If an assertion about unobservable phenomena occurs in the epistemic context of what is to be believed, accepted, asserted, or inferred, we bracket the assertion and append to it the operator "is empirically adequate." If, then, it is asked why a theory is successful, the answer need not be that the theory is true or bears any similarity to a true theory, but merely that it is empirically adequate.

It may be objected that empirical adequacy just *is* success, so it cannot *explain* success. But this is incorrect on two counts. First, empirical adequacy generalizes success; it is not identifiable with any achievable record of specific successes, however large. In this respect, a claim of empirical adequacy exceeds any possible body of evidence for it and assumes a hypothetical status, subject to confirmation and disconfirmation much as the theory itself is ordinarily thought to be. Second, an explanation of a theory's success is not simply an explanation of empirical

---

17. See van Fraassen, *The Scientific Image.* Van Fraassen favors a semantic analysis of theories, and, in presenting that analysis, sometimes *identifies* a theory with its class of empirical models. This would *reduce* a theory to its empirical consequences, rather than replace theoretical claims by something weaker. But as van Fraassen holds that theoretical claims have unknowable truth value, his model-theoretic definition of theory is not strictly tenable. Statements of unknowable truth value can be identified neither with empirical statements of knowable truth value nor with empirical models devoid of truth value. This is just as well, as *which* empirical systems model a theory is a serious question of application, not decidable definitionally.

findings that bear on the theory. As urged above, it is additionally pertinent to ask of a theory why *it* yields what is empirically found. Empirical adequacy may then be cited as the attribute that enables it to do this.

It is to be admitted that van Fraassen himself does not emphasize the role of empirical adequacy in explanations of success. His main concern is to show that what is observed never warrants asserting more than a theory's empirical adequacy, never warrants asserting the theory itself. When it comes to explaining success, van Fraassen is more inclined to give an evolutionary account or to insist on a pragmatic analysis of explanation, approaches already discussed. This is why I spoke of "associating" the view that truth is replaceable in explanation with van Fraassen, rather than attributing it to him explicitly. But van Fraassen does frequently cite empirical adequacy in explanatory contexts as well as epistemic ones, always as an adequate alternative to truth or assertion.[18] And the view that empirical adequacy is all we need to account for whatever needs explaining about a theory's performance is, in any case, worthy of consideration as an opponent of the position I take, especially as van Fraassen's preferred alternatives have been discredited earlier.

However, as an explanation either of what is observed or of how a theory manages to predict what is observed, empirical adequacy leaves a lot to be desired. Most important, empirical adequacy is an attribute that itself cries out for explanation. Even if we do not deny it explanatory resources, it still leaves the sort of explanatory lacuna lately found unacceptable. "Why is a theory empirically adequate (insofar as evidence supports the conjecture that it is), unless it is true or in some way like what is true?" is the question to press. What is it about the theoretical structures it posits that enables them to yield evidence of its empirical adequacy—an attribute wholly independent of those structures, definable without reference to them—if not truth or similarity thereto? Once we are prepared to impute truth to a theory, such a further question occasions no embarrassment. A theory is true either because of the truth of some deeper theory to which it reduces, or, if it is fundamental, because that is the way of the world. But imputing empirical adequacy generates a further question that cannot be thus dismissed, for *it is a question that truth would answer.* The truth of a theory would explain its empirical adequacy. So we are entitled to ask, "If not truth, what?", and to be dissatisfied with silence when truth is available.[19]

As an explanation simply of the observed facts, empirical adequacy is, if anything, less appealing. Generalizations do not explain their instances. On the contrary, the apparent satisfaction of a generalization is more remarkable, and more an inducement to seek an underlying structure or a deeper principle, than the individual case. We theorize because there is a difference between generalizing, predicting, sorting, organizing, systematizing, or whatever, our observations, and

18. For example, van Fraassen, *The Scientific Image*, pp. 19–20, 80–82.

19. Of course, we could be dissatisfied with truth, too; sometimes, no explanation is better than the only explanation available. But the adequacy of truth as an explanation is a separate question already addressed.

explaining them. Perhaps the first impulse to theorize was patterns remarked in celestial motions, patterns suggestive of differences in the objects that could not be visually discerned. The recognition of patterns in observations is a major scientific achievement. But it is an achievement that identifies a theoretical task; it is not an explanatory achievement.

It might be supposed that generalizing observed facts explains them if the generalizations are *nomic*. That is, the hypothesis that a generalization is a law of nature explains its instances. Now, to be sure, this hypothesis is logically stronger than the facts themselves; it does not merely restate them. And it implies them, so that if it is true, then the observations are to be expected. But I do not think that this is all we require of an explanation. We want some understanding of how the facts are produced, some process or agency that establishes them. For this we need to place the law within a broader theoretical context. The mere attribution of nomic status, without any theoretical basis, is gratuitous; it imparts no further understanding. Responding to explanatory challenges simply by elevating what is observed to the status of law is not explaining, but *refusing* to explain. At best, it is to claim that no explanation is necessary. This claim could, admittedly, be correct. But to judge it so requires a theoretical basis.

Fine goes further than van Fraassen toward theory in response to the call to explain success, although it is not clear that Fine recognizes in that call anything more than explanation of predicted phenomena.[20] I understand Fine to hold that theories explain by themselves, and that there is nothing about their ability to do so that in turn needs explaining. But for one who does raise this further question, Fine offers the following prescription: Avoid hypothesizing truth or any connection thereto of a successful theory, in favor of hypothesizing that the world operates *as if* the theory were true; prefix an "as if" operator to theoretical statements used in explanation. If everything happens just as it would were the theory true, then of course we should expect the theory to be successful. So Fine decides that this maneuver suffices to account for any success that we may be inclined to invoke truth to explain.

Further, Fine and, analogously, van Fraassen, make much of the fact that theories imply the "purged" or truth-excised versions of themselves that result from applying the preferred operator, be it "as if" or "is empirically adequate." Because of this logical relation, any observations taken to confirm a theory, or demonstrate its successfulness, must equally be taken to confirm its purged version. The purged version is epistemically weaker, closer to what is given in experience; therefore a lesser degree of error is risked in inferring it than in inferring the original. Since any evidence for a theory is, equally, evidence for its purged version, and the purged version carries less epistemic commitment, the purged ver-

---

20. The pertinent view of Fine's is developed in a series of essays: "The Natural Ontological Attitude," in Leplin, *Scientific Realism*; "And Not Anti-realism Either," *Nous*, 18 (1984):51–65; and "Unnatural Attitudes: Realist and Instrumentalist Attachments to Science," *Mind*, 95 (1986):149–179. The first two of these are available in Fine, *The Shaky Game; Einstein, Realism, and the Quantum Theory* (Chicago: University of Chicago Press, 1986). The third is most important for present purposes.

sion must be better justified than the original. Having supposed that the purged version has the same explanatory resources as the theory, Fine therefore concludes that the purged version is a better explanation.

There is a small irony to note here. Judging a theory and its purged version only as purported explanations without regard to their acceptability, the purged version cannot very well be better. For it can contain no explanatory resources absent from a theory that implies it. What makes it better is only its differential justificatory stature. Why should this make it better *as an explanation*, unless explanations *need* justification? And why should explanations need justification— that is, need it to excel as explanations—if it is not the case that only truth or some connection thereto explains? Justification is a truth indicator; its importance is to direct conviction. In contending that the purged version explains better, Fine appears to concede that explanations must be true. To be consistent, he should say only that the purged version is better justified, not that it is a better explanation. Perhaps that is what he *intends*, but what he *says* is significant as an indication of the intuitive force of the role of truth in explanation.[21]

The point can be made as well by shifting perspective from the purged version to the theory. What makes the theory *worse* as an explanation? Evidently, the answer is that the theory carries avoidable epistemic commitments; it posits unobservable entities, perhaps. Why are such commitments avoidable? They are alleged to add no explanatory resources to what is available without them in the purged version. A methodological principle, "Ockham's razor," counsels the excision of superfluous entities. Why should one avoid commitment to such entities? The answer is that their superfluousness implies that we are not in a position to endorse them; they lack justification, or lack it relative to the versions from which they are purged. In effect, then, the original theory is a worse explanation because we are in no position, or in less of a position, to affirm its truth. So again, truth must matter in explanation.

The criticism put to van Fraassen can be tried on Fine. Why does everything happen as if the theory is true? The theory's truth would explain this, so it would seem a legitimate query. But Fine cannot allow this query, because he disallows truth and offers nothing in its place. Indeed, a major point of his philosophical position, the "natural ontological attitude," is that recourse to any surrogate for truth is illegitimate. So he leaves an explanatory lacuna. He is welcome to argue that the lacuna is real, that the as-if version is all we can ever know or justify. But he is not entitled to claim that this version is as good an explanation as truth or some semblance thereto would provide, let alone a better one.

I do fault Fine on this point. However, it is noteworthy that he gives us more of an explanation than does van Fraassen, who disallows any reference to unob-

---

21. Fine's argument in "Unnatural Attitudes" turns on the thesis, for which he claims the status of metatheorem, that "*if the phenomena to be explained are not realist-laden, then to every good realist explanation there corresponds a better instrumentalist one*" (ital. in original). By an "instrumentalist" explanation, here, he means one that excises the truth or assertion of a theory by prefixing the "as if" operator.

servables. Fine's proposal is to say of the world itself, not just its observationally manifestable features, but its allegedly unknowable structure as well, that it has the property of producing just those effects that would be produced were that unknown structure to be as the theory represents. The proposal posits a structure beyond the appearances as the cause of the appearances; van Fraassen only tells us what the appearances are—the ones we get by using the theory. Fine's proposal at least acknowledges that just generalizing observations does not achieve as good an explanation as a theory's truth would provide of what we observe or of the theory's ability to predict it.

The view that emerges from Fine's proposal provides that the world has a "deep structure," a structure not experientially accessible, but causally responsible for what we experience, which theories attempt to represent. No representation of it is endorsable, however. The best that can be said for any representation is that the actual structure produces the observable effects that it would produce were that representation true of it. The explanation of the success of any theory, however great and exceptionless, is that the actual structure of the world operates at the experiential level as if the theory represented it correctly. Whether the theory does represent it correctly is, in principle, undecidable.

I call this view "surrealism" (for "surrogate realism") to contrast it with a *scientific realism* that regards the accuracy of theoretical representations as in principle amenable to epistemic evaluation.[22] There is an *ontological realism* with which surrealism is quite compatible. It holds that there are deep structural facts about the world to be represented, and that theoretical statements are definitively true or false, depending on what those facts are. It says nothing about the possibility of learning such facts or of determining whether or not any theoretical statement is true; this is the business of *epistemic realism*, with which surrealism is incompatible. A minimal epistemic realism holds that such discoveries are possible, that there are possible empirical conditions, realizable in principle, under which we would be justified in judging some deep structural statements to be true, or, at least, partially so. More ambitious epistemic realisms go so far as to endorse some of the theoretical structures of extant science. I am going to argue that the success of science is not adequately explained without some epistemic realism. As a step in that argument, surrealism must be dispatched.

The skepticism embodied in surrealism has several possible motivations, most obviously, the underdetermination thesis, which would allow us to deny theoretical statements but never to assert them. But it is the explanatory pretensions of surrealism that pose the immediate challenge to epistemic realism.[23] If an adequate or unimprovable explanation of a theory's success does not require imputing truth to the theory, then a major reason for supposing truth to be discoverable is lost. Certainly, a major motivation for epistemic realism is the conviction that unless there is some truth to a theory, there must be some limit to how successful

22. See J. Leplin, "Surrealism," *Mind*, 97 (1988):519–524.

23. Let me note again that it is at most the ability of surrealism to provide an explanatory alternative to realism that Fine is committed to; he does not advocate surrealism itself.

the theory can be. If not impossible, it is at least improbable that a theory wholly misrepresentative of the structure of the world will be highly successful to exacting standards. It would require a colossal coincidence for the theory to work out exactly right so far as we can tell, while being completely wrong at the level of deep structure. Sufficient success, then, would warrant imputing some truth to the theory, albeit fallibly. For success not to provide such warrant would be for it to occur as likely as not by coincidence, for it to be unexplainable.[24] Surrealism preempts this line of reasoning. By proposing to explain success without truth, it leaves the inference from the success of a theory to its truth without warrant.

I therefore wish to challenge the explanatory pretensions of surrealism further, by arguing that not only does surrealism leave an unnecessary explanatory lacuna, but also *such explanation as it does provide is actually parasitic on realist explanation.* Suppose we press for specificity as to what it means to assert that the world acts "as if" a theory is true. What implications are we to suppose such behavior to have at an experientially accessible level? The implications must be the same as those of the theory's truth, and what are they? Evidently, we are supposed to assume that experience will bear out the theory's predictions. For if the predictions of a true theory need not be borne out, then asserting that the world behaves as if the theory is true does not explain why the theory's predictions *are* borne out; it does not explain the theory's success. Therefore, if it is to explain success, if it is to compete with realism as an explanation, surrealism must *presuppose* that theoretical truth will be manifested in experience. But that presupposition is precisely the explanation realism gives of success. The realist intuition is that if we have latched onto the right theoretical structure, our predictions, based on this structure, will conform to experience. Thus, surrealism is explanatory only in virtue of presupposing realist explanation.

It is not just that in order that surrealism be explanatory, realism must be as well. In promoting its own explanation as an alternative, surrealism need not deny that realism has an explanation to offer. The problem is, rather, that the particular explanation realism gives is actually embodied, by implication, in surrealism's explanation.

Surrealism tries to strike a middle ground between the success it would explain and the realism it would supplant. It plays the role of intermediary by proposing a theory's *reliability at anticipating experience* as an alternative to its truth in the explanation of its success. But there is no distinctive middle position to be staked out, and surrealism shifts, under analysis, into its opposition, if not its object. If

---

24. Van Fraassen argues that coincidences need not be unexplainable (*The Scientific Image*, p. 25). He even cites Aristotle as authority (*Physics* II). What Aristotle acknowledges, but van Fraassen does not, is that an explanation of coinciding conditions need not constitute an explanation of their coincidence. Aristotle emphasizes that it is in examples involving intention or purposive agency that once coinciding conditions are explained, no (further) explanation of their coincidence may be required. But if a theory is successful by coincidence, we do not have independent explanations of coinciding conditions that collectively obviate explanation of their coincidence. We have *only* an explanation of how predictions, which happen to be correct, were made, which, I have argued, fails to address the question of why the theory is successful.

reliability just *is* empirical success, then of course theoretical truth is not invoked. But also, then, surrealism becomes redundant with its explanatory object, and offers no explanation; it collapses into mere empirical adequacy, which just generalizes what is to be explained. Attempting to avoid the collapse by adding that the generalizations are nomic is not, I have argued, explanatory. If, instead, reliability is not success, but a property of a theoretical structure, then the explanation it gives of success is, implicitly and by implication, the very explanation realism gives.

A theoretical structure could be reliable without being true of the world; the world could operate as if a theory represented it, although the theory gets the world completely wrong. For there could be different theoretical structures indistinguishable at the level of experience. This observation is a challenge with which realism must contend if its explanation of success is to be defensible. It is essentially the challenge of underdetermination, once again. But this admittedly pertinent observation exhausts the resources of surrealism; it cannot be parlayed into an independent explanation of success to compete with the explanation realism gives.

Surrogates for Truth

I have considered empirical adequacy and reliability of theories as alternatives to their truth in the explanation of their success. Neither of these properties is intended to supplant truth as a concept. On the contrary, to champion either property as the ideal of scientific achievement is to assume a self-consciously *skeptical* position with respect to the scope of scientific knowledge. It is to concede— indeed, to insist—that truth is *not* achievable (except, perhaps, by unknowing happenstance). If, by contrast, we count *epistemic surrogates* for truth, analyses of the concept of truth that purchase its epistemic accessibility by excising its metaphysical commitments, then the list of alternatives to consider extends. All manner of truth-surrogates have been proposed as what science *really* aims for, all just happening to have the convenient feature that the possibility that successful theories lack them does not arise.

Warranted assertability is a favorite; to declare a theory true is alleged to reduce to declaring that the evidence warrants asserting it. Thus, any metaphysical worry about the possibility that we are warranted in asserting a false theory is obviated. Pragmatic and instrumental utilities are other proposed truth-surrogates; there can then be no danger of being misled into endorsing a false theory by its usefulness. Freedom from grounds for doubt has been tried, eliminating the risk that a theory's inaccuracy will fail to be manifested in some experience that gives us cause to question it. More subtle accounts have been proposed by Hilary Putnam and Brian Ellis.[25] In general, the candidates differ according as conceptions of what we go on in evaluating theories differ.

25. Hilary Putnam, *Reason, Truth, and History* (Cambridge: Cambridge University Press, 1981), and Brian Ellis, *Rational Belief Systems* (Oxford: Basil Blackwell, 1979).

I shall not trouble over the explanatory resources of such surrogate concepts. For my money, all of them constitute methodologically otiose attempts to claim epistemic victory by changing the rules of engagement. Rather than figuring out how, if at all, we could be justified in claiming to have discovered theoretical truth, one simply redefines 'truth' so that the kind of evidence science ordinarily achieves automatically qualifies as discovering truth. Happily, the epistemic goal toward which inquiry has always been directed is discovered to coincide with the very epistemic conditions that inquiry happens to achieve. It may make political sense to get out of a tough war by declaring victory and going home, but it is not respectable intellectually. A reductive analysis of truth is not necessarily illegitimate, but the standard of its correctness must be independent of the convenience of its result. It is blatantly ad hoc to make the achievements of science definitive of its objectives; one might as well claim to solve a problem by erasing the blackboard.

All truth-surrogates designed for epistemic accessibility share in common the disadvantage of requiring relativization to a particular stage in the growth of knowledge. What we are warranted in asserting today we may not be tomorrow; today's utility is tomorrow's curiosity. Truth does not operate this way, and any analysis that so portrays it is not an analysis of truth. It is fundamental to the concept of truth that truth be stable under the vicissitudes of human credulity and convenience. It did not become true that light is a wave with the discovery of diffraction, and it did not cease being true that light is a wave with the discovery of photoelectricity. Like it or not, the concept of truth is unabashedly metaphysical. The true structure of the world could be other than experience suggests, other, even, than experience could suggest. There is no inconsistency in the supposition that the inferences for which evolution has equipped us do not reveal deep structure, that evolution has not selected for accuracy of representation at that level. There can be no a priori exclusion of the possibility that we err *systematically*, that the true structure of the world might never be distinguished by the evidence available to us, even in principle or in the ideal limit of human inquiry.

The challenge of epistemic realism is to show that there is, or could be, good reason to reject these possibilities; it must not define them away. Truth is accuracy of representation of an independent world—a world that, while it includes us and our acts of representing it, decides the accuracy of our representations of it and is not constructed by them. Knowledge of anything less does not establish epistemic realism.

The Metaphysical Import of Truth

The discussion to this point invites the conclusion that there is no getting around truth, or some connection to it, if the success of theories is to be fully explained, that nothing but truth is ultimately explanatory. I think this conclusion is correct in spirit, and establishes the right direction for interpreting the epistemic standing of science. But there are important disclaimers to make to the letter of this conclusion, and I shall complete this chapter by clarifying the explanatory connection of truth with success on which the continuation of this book depends.

The case I have made for truth does not imply that success *is* explainable; success might have no explanation although it needs one, and might neither have nor need one. About the former possibility, I have said only that we should come to it, if at all, out of desperation or on the strength of positive argument; the weakness of some purported explanations and the reminder that success could be chance do not lead to it. As to the latter possibility, I have noted some forms of success that do not properly invite imputations of truth, because it is perfectly clear how they might otherwise be achieved—by persistence or by trial and error, for example. However, persistence and trial and error do not really explain *theoretical* success, or so I have argued. They explain *our* success at finding formulas or hypotheses, occasionally theories, that work, without explaining *why* they work. Nor does the refinement of such methods, through the constraints and checks we evolve to avoid pitfalls we learn to recognize, make such an explanatory difference. Striving for objectivity through controlled and blind experimentation decreases the likelihood that a hypothesis will pass muster, but does not explain why one that does is successful.

To such examples must now be added some cases of theoretical success that do not properly raise questions of truth. For it is by no means the case that the success of a theory is always or necessarily to be explained, if at all, by imputing truth to the theory, and we must circumscribe the range of cases to which the relation of truth to success will apply.

From Newton's theory of gravitation, one may deduce the astronomical facts encapsulated under Kepler's three laws of planetary motion, and the facts of terrestrial motion covered by Galileo's law of falling bodies and by his law of pendular motion, among others. It is not these laws themselves that are strictly deducible, for they are not quite correct; rather, the observable facts that instantiate the laws are predictable directly from Newton's theory. The ability of Newton's theory to unify such diverse phenomena as celestial and terrestrial motions is a major success.

From Einstein's special theory of relativity, the null result of the Michelson-Morley experiment is deducible. The fact, discovered by Albert A. Michelson, that the difference, if any, in the round-trip travel times of light along different paths of an interferometer is independent of the orientation of the interferometer with respect to the direction of the earth's orbital motion is a direct consequence of relativity theory. Michelson's result was historically important as a basis for preferring relativity to competing theories that extended the laws of electromagnetism to moving bodies, or moving systems of reference, theories for which Michelson's result was persistently anomalous.[26]

In neither of these examples does the explanation of the success in question require supposing that the theory achieving it is true, however. The reason is that

---

26. For extended analysis of this case, see J. Leplin, "The Concept of an *Ad Hoc* Hypothesis," *Studies in History and Philosophy of Science*, 5 (1975):309–345.

generalizations of the predicted observations were explicitly and essentially involved in the reasoning by which the theories were developed. Newton used Kepler's laws, for example, as assumptions in deducing the form of his law of gravitation. Einstein assumed that the velocity of light is independent of the choice of coordinate system with respect to which velocities are measured. Whether true or not, a theory that makes this assumption is bound to predict that no such variation in velocity as Michelson sought to measure will be found. In short, it is no mystery suggestive of a theory's truth that predictions instantiating the very assumptions on which the theory is based are borne out. It is still necessary, of course, that a theory be true if it is to explain the observed phenomena. But how the theory manages to predict these phenomena does not require its truth for explanation, if the phenomena are assumed in constructing the theory. If success is to cast some credibility on a theory beyond that represented just by knowledge of the observational facts, the theory must enjoy some measure of independence from these facts.

The scope and nature of that independence are problematic, in part because dependence of some sort is necessary if the facts are to be obtained as predictions of the theory. As prediction is deductive, there must be a logical connection between a theory's laws or hypotheses and descriptions of the facts it predicts. The independence needed will pertain more to the structure of the theory's provenance than to the final form of its laws. But clearly, some predictive success will be obtainable automatically, and will not argue for a theory's truth.

Still another disclaimer must be issued regarding the explanatory resources of truth with respect to theoretical success. I have argued both that explanations must be true and that truth is explanatory. Specifically, I have urged that the truth of a theory would suffice to explain the success of its predictions. On behalf of the legitimacy of this contention, I might simply have pointed out that it is conceded even by philosophers who deny that success betokens truth. But I claimed the prerogative of assuming this sufficiency without argument or authority, on the basis of what we regard as constituting a theory, on the basis, that is, of the essential explanatory function imputed to theories. However, when it comes to further forms of success, beyond producing the right predictions, the explanatory resources of truth cannot be thus assumed. In addition to the qualification just noted, that truth may not be required to explain how a theory manages to produce the right predictions, I must consider how, when it is required, truth manages to explain other forms of theoretical achievement. For it is a simple matter to envision situations in which a true theory could be of limited utility. Perhaps its predictions are correct but it makes very few of them, and it makes no contribution to the solution of outstanding problems that a theory of its type would be expected to solve. A theory might be judged unsuccessful, on balance, despite being true. How does being true help to explain forms of success that a true theory need not achieve?

Beyond the correctness of such predictions as a theory does afford, truth explains success only conditionally on assumptions that, if deeply entrenched in scientific method and indirectly supported by vast experience, are nonetheless metaphysical in character. This is because the concept of truth itself is unavoidably metaphysical, as I have argued in attempting to discredit surrogates for it. Truth, as ordinarily conceived, is a condition that could remain concealed despite our best efforts to elicit it. There is no inconsistency in supposing that at the level of deep structure, of the entities and processes ultimately responsible for our experience, the world is other than our best supported theories represent it. Perhaps our evolutionary history, in equipping us to flourish in an environment of middle-sized objects, has selected against the sort of attributes we would require to read microstructure or superstructure correctly. Perhaps we are systematically misled in our efforts at theorizing, and are unequipped to correct the error. Perhaps it is impossible for us to acquire evidence against certain false beliefs, because these beliefs do not differ from true ones at an experientially accessible level. According to the natural conception of truth as representation of a theory-independent world, such possibilities are always entertainable.

Therefore, judgments about what a true account of the world would be like, however supported by scientific experience, contain a metaphysical component that transcends experience. And such judgments are unavoidable in invoking the truth of a theory to explain its success. Suppose that a theory is well confirmed by a body of evidence of a certain kind, say, by observations of the atomic spectra of hydrogen gas. One would expect the theory to provide a basis for analyzing the structure of helium as well. Experiments with helium and ensuing chemical elements become a legitimate test of the theory, and if it fails, its correctness, even for the case of hydrogen where the results were confirmatory, is called into question. Likewise, observations of stellar spectra are expected to be compatible with the theory. In treating such additional domains of evidence as tests of the theory, it is assumed that the atomic structures of different chemical elements are related in basic ways, that there are common structural principles that a correct theory for the simplest element will reveal. And it is assumed that extraterrestrial light sources, whether or not composed of the same elements as terrestrial sources, are composed of elements that obey the same structural principles.

These assumptions have no basis in logic. And although borne out by experience, they are not *based on* experience, for the failure of predictions in such new domains would discredit not the assumptions, but the theory to which they are applied. They are metaphysical assumptions in that they involve deep-structural depictions of the world that transcend experience, and they serve as presuppositions in the interpretation of experience. They are instances of the general assumptions that, at a sufficiently fundamental level, the laws of nature are the same everywhere and are manageably limited in number and complexity. Theoretical science in our time could hardly operate without such assumptions, and the success of methods that presuppose them is empirical support of a kind. But they are unabashedly metaphysical.

The contemporary drive for unification of the fundamental forces of nature is a dramatic manifestation of the entrenchment of the assumption of such simplicities in scientific method. Virtually every theoretical physicist expects gravity to be unifiable with the other three forces—expects, that is, that quantum theory will be applicable to gravity—even as he despairs of existing efforts at unification. Before acquiescing in the ultimate independence of the relativistic theory of gravitation from quantum theory, he will reconsider the correctness of each theory with respect to its established domain of successful application. This attitude is a strong argument for the metaphysical status of the assumption that the world is simple.

Thus, if we expect the truth of a theory to explain the success of the theory at anticipating experiences of a kind not involved in its development or original domain of application, if we expect successful theories to be successfully *extendable* and we count failure in this further obligation against their *correctness*, then we commit ourselves to a certain metaphysical picture of the deep structure of the world. This picture has no a priori warrant. Yet I shall argue that this picture is ineluctable if we are to explain some forms of theoretical success, and that its ability to provide such explanation can warrant imputing a measure of truth to theories incorporating it.

2

Conceptions of Novelty

The History of the Independence Requirement

The idea that a prediction can claim some measure of independence from the theory that yields it is not new. It was connected in the nineteenth century with "the method of hypothesis," a precursor of hypothetico-deductivism. These methodologies are alike in admitting into scientific theories explanatory hypotheses that cannot be induced from observations. Both methodologies depict scientific inference as the deductive explanation and prediction of empirical results from hypotheses. But unlike hypothetico-deductivism, the method of hypothesis is primarily a methodology of confirmation; it supposes that the ability to explain and predict bespeaks truth. The emphasis is not on testing at all. The point is not to confront hypotheses with empirical results that they should, if true, predict accurately, but to evaluate hypotheses on the basis of their explanatory and predictive resources. Hypotheses are to be valued in the measure that they provide explanations and predictions of important empirical phenomena. The failure to provide these in some instances can be offset by successes elsewhere, so that, on balance, the hypothesis excels. Explanatory success on balance, especially relative to alternatives, is reckoned a suitable basis for acceptance.

But what are the sorts of empirical results whose explanation or prediction would qualify a hypothesis for acceptance? Nineteenth-century advocates of the method of hypothesis—William Whewell, in particular—were not so naive as to ignore the possibility that false hypotheses would predict correctly.[1] The issue, rather, was whether sufficient constraints could be formulated for explanation and prediction to ensure that explanatory and predictive success would not accrue, or would be unlikely enough to accrue, to false hypotheses.

John Stuart Mill, following Newton a century earlier, thought not; Mill decided that an unacceptably large potential for misdirected credence was ineliminable on this method. On the other hand, he held to the ability of inductive methods to establish some hypotheses directly. That is, he thought that observations could be generalized and connected so as to construct a hypothesis out of them, in a process distinct from the predictive or explanatory use of hypotheses already formulated. This process was thought to be at once creative and confirmatory, the method of constructing hypotheses itself attesting to their veracity. Newton's inference from observations with prisms to the composition of white light was the leading example. Thus, Mill rejected Whewell's method of hypothesis, only to advocate an alternative "inductivist" methodology of confirmation in its place. Hypotheses that could not be induced from observations were yet potentially affirmable for Whewell, but were methodologically illegitimate for Mill.

A good example of such a hypothesis is Charles Darwin's controversial contention, in *The Origin of Species*, that the processes of natural selection were capable of accounting for variation in plant and animal life. It was not so much Darwin's specific claims about the existence of selective processes, and about the heritability of traits bearing differentially on reproductive survival, that made his evolutionary theory controversial. What raised methodological concerns was his insistence that selection could fully account for the variation in genotypic traits and the geographic distribution of species.[2] Darwin was criticized, in the 1830s, by William Hopkins and Richard Owen for advancing speculative hypotheses that could not be supported by the available evidence. There was no evidential basis, the critics charged, for preferring evolutionary hypotheses over long-standing alternative views, particularly the belief that species were closed under reproduction. The explanatory promise of evolutionary hypotheses was not denied; rather, it was held to be an illegitimate basis for promoting them in the absence of observations of plant and animal life sufficient to prove that natural selection could produce speciation. Darwin was criticized for relying on explanatory power as the primary argument for his theory.

Specifically, Darwin's approach was accused of being unscientific, of relying

---

1. Less sophisticated presentations of the method of hypothesis were popular in the eighteenth century. Huygens espoused the method in defense of his wave theory of light, and Hooke, Boyle, and Descartes were other early advocates of it. Ronald Giere traces the method back to the interpretation of astronomical models in ancient Greek science. See his essay "Testing Theoretical Hypotheses," in John Earman, ed., *Testing Scientific Theories*, *Minnesota Studies in the Philosophy of Science*, vol. 10 (Minneapolis: University of Minnesota Press, 1983). Giere thereby follows Pierre Duhem, whose *To Save the Phenomena* (Chicago: University of Chicago Press, 1969) interprets Greek science in terms of the method.

2. See Philip Kitcher, *The Advancement of Science* (New York: Oxford University Press, 1993), chap. 2, sec. 3.

on faith rather than reason, of failing to provide "deductions from the evidence," of amounting to "mere hypothesizing."[3] The language used in mounting the attack was very close to that used in complaints raised against the wave theory of light during the same period, and Darwin responded by explicitly citing the explanatory success of the wave theory as a defense of his method. He argued that his detractors, in limiting themselves to what could be established directly from the observational evidence, were simply incapable of addressing a range of biological questions to which evolution proposed promising answers. He offered the testing and development of these answers as a new agenda for biological research, creating a methodological struggle for the resources of the discipline.

The wave, or undulatory, picture of the composition of light, to which Darwin appealed, provides a prime example of a hypothesis that could not be induced from observations, as Mill required, but was potentially affirmable on Whewell's methodology. Theories of light divided the scientific community of the early nineteenth century along methodological as well as physical lines. The wave theory was understood to require the hypothesis of an all-pervasive ether medium for the transmission of light waves, and such a hypothesis failed to meet the Newtonian standard of inducibility from observations. Mill dismissed the hypothesis for this reason, while Whewell defended it on the basis of its explanatory potential. Contributors to the wave theory (Thomas Young, who was responsible for its nineteenth-century resurgence; and Augustin Fresnel, who developed it mathematically) sided methodologically with Whewell, while the wave theory's detractors (Pierre Simon Laplace, George Airy, Siméon Denis Poisson, and Jean Baptiste Biot) represented Mill's position.

The rise in the fortunes of the wave theory, at the expense of particle theories based on the application of Newton's mechanical laws to corpuscles emanating from a luminous body, evidently depended on a concomitant shift in the fortunes of the contending methodologies. Inductivism, tracing its authority to the success of Newtonian science, was dominant at the beginning of the nineteenth century, and pervaded early criticism of the work of Young. By the late 1830s, deductivist arguments, appealing to the explanatory success of the ether hypothesis, were common in explanations of the relative superiority of the wave theory, as they were in debates over method in biology.

As may be expected, however, the actual physical reasoning used by scientists on both sides of the question of light's composition is considerably more complicated than capsule summaries of the alternative methodologies would suggest. In particular, wave theorists did not limit their case to the explanatory resources of the ether, but adduced numerous independent criticisms of particle theories. These concerned primarily the implausibility of assumptions needed to apply Newton's laws to the phenomenon of diffraction, the dispersal of light into the geometric shadow of an aperture or of an opaque barrier.

Because arguments used to defend the wave theory were more comprehensive than the method of hypothesis represents, the case has been made that no real

---

3. Kitcher, *The Advancement of Science*, chap. 2, sec. 6.

methodological competition was involved in the comparison of wave and particle theories. Because induction from observations must have been used by wave theorists in faulting their rivals' analysis of diffraction, for example, it is denied that the shift to wave theory constitutes a methodological change away from inductivism.[4] Hence, we can toe a monomethodological line in this case, declining, despite the success of the wave theory, to acknowledge a role for explanatory power as such in the confirmation of hypotheses. And there is the added advantage of forestalling the potential challenge to scientific objectivity implicit in the instability of methods of evaluation.

It seems to me, however, that this analysis cannot make sense of the very significant role, in the defense of the wave theory, of new observational phenomena—such as spherical diffraction—that were readily explainable on a wave picture, but were initially mysterious and requiring of artificial embellishments if light is regarded as corpuscular. Such phenomena made a real difference to the credibility of the wave theory, a difference that the monomethodologist is hard pressed to account for. In particular, it is implausible to represent their role as merely that of maintaining consistency, serving to bear out theoretical allegiances already in place.[5] To account for this difference requires treating the ability to explain new phenomena as important evidence in favor of a theory, which is improper in Mill's methodology.

If we do not allow methodological discord in this case, and, with it, the prospect of change in prevailing methodological norms, we cannot understand the reasoning of the scientists involved. Both the wave theory and the particle theory had their proponents, and each took the evidence, or some proper part of it, to support the theory he preferred. Biot and Poisson thought that rectilinear propagation, refraction, and dispersion supported a particle theory; Fresnel thought that the independence of the speed of light from its source, interference, and diffraction supported a wave theory. But each knew how the theory he opposed could be made to handle, at least qualitatively, the phenomena he was most inclined to cite in support of the theory he advocated, even if he found this accommodation objectionable. If diffraction was particularly problematic for the particle theory, polarization proved similarly troublesome to the wave theory. A disinterested view might hold, for a time at least, that the evidence, on balance, supported both theories about equally. That the parties involved did not take this neutral view requires attributing to them some differences as to the kinds of evidence that matter; specifically, as to the weight of explanatory advantages in assessing theories. In general, where scientists draw conflicting theoretical conclusions from a

---

4. See Peter Achinstein, *Particles and Waves* (Oxford: Oxford University Press, 1991).

5. This is Achinstein's treatment of the importance of new explanatory success of the wave theory. Pressed to explain the rise of the wave theory, and unwilling to credit explanatory differences whose influence would have depended on a shift to the method of hypothesis, Achinstein reverts to the wave theorists' original position, which he sees new optical phenomena as simply reinforcing. This is typical Bayesian evasiveness, relying on subjective assignments of prior probabilities that need not be defended. The use of Bayesian principles to account for the role of new explanatory successes in the confirmation of theories is addressed in this chapter under "Bayesian Confirmation."

common body of relevant evidence, it would seem unavoidable that some methodological precepts distinguish them.

Certainly by the middle of the nineteenth century, prevailing methodology—partly as a response to the success of the wave theory, and partly through a reevaluation of Newton's methods, which, after all, countenanced the very abstract idea of universal gravitation despite Newton's official opposition to hypotheses—provided a major role for explanation in the evaluation of hypotheses. The method of hypothesis, like the inductivism that it supplanted, provided for the warranted acceptance of hypotheses; only the method of hypothesis allowed a wider range of hypotheses to be warranted by diversifying the allowable forms of warrant.

It was left to Karl Popper, early in the twentieth century, to dismiss confirmational methodology altogether, settling for the purely disconfirmational, or "falsificationist," component of hypothetico-deductivism as the sole form of epistemic inference appropriate to science. Popper's rejection of the project of confirmation theory is supposed to emerge from an attack on inductive inference as systematically incapable of warranting belief. It seems clear, however, that no sweeping dismissal of inductive inference is available to Popper. For he wishes to uphold the falsifiability of hypotheses, and is bound to accede to Pierre Duhem's admonition that some positive epistemic commitments are logically required as premises in any valid deduction of the falsity of a hypothesis. Such auxiliary commitments could only be inductively based.

Popper's official response to this problem is to deny that these commitments are "based," or epistemically grounded, at all. Rather, he treats assumptions needed to ensure that it is on the falsity of hypotheses that predictive failures are to be blamed as "conventions," meaning that there is something ineliminably arbitrary about accepting them. He then elaborates a theory to the effect that scientific method in general is conventional. This outcome is quite unsatisfactory for Popper, in that his agenda is to elevate the epistemic credentials of science above those of all other forms of inquiry. To found the empirical knowledge that certain hypotheses are false on epistemically arbitrary conventions does not seem much of an improvement on a foundation requiring dubious modes of inference.

What seems really to be going on is that Popper has precluded the viability of confirmation from the beginning by establishing falsifiability as the defining feature of the scientific status of hypotheses. It is not so much that hypotheses are unconfirmable because of problems with induction, as that their status *as scientific hypotheses* renders them unconfirmable. To be a genuine scientific hypothesis in the first place, a statement is required to be falsifiable, and the more falsifiable the better. The easier a statement is to falsify—the more possible empirical situations there are that conflict with it—the more "information content" it has, and, thus, the more "scientific" it becomes. In effect, scientific status is measured by *audacity*, the audacity of exposure to ready refutation. But the prospects for survival are low according as the risk of refutation is high. So conversely, scientific stature is won only to the extent that confirmability is lost. By meeting Popper's condition for scientific status, a hypothesis automatically resists confirmation, whatever is decided about the legitimacy of induction.[6]

Contemporary opponents of epistemic realism have no call to acquiesce in Popper's renunciation of induction. They are free to decry the inferential sterility of his falsificationist methodology, and to recognize a variety of dimensions of appraisal for hypotheses. They may admit epistemically grounded preferences among hypotheses. More "fallibilists" than falsificationists, they may insist on the defeasibility of all knowledge claims, not as a conceptual requirement on "knowledge," but as a fact of scientific life. So long as they do not count truth among the dimensions of appraisal for hypotheses, they continue the essential Popperian legacy of disbelief in theory.

Now, how does a sophisticated apologist for the method of hypothesis, like Whewell, accommodate the concerns of fallibilism, and prevent false hypotheses from surviving the tests that are to warrant belief? He insists on the *independence* of probative predictive successes. This is really what Whewell's "consilience of inductions" amounts to. "Consilience of inductions" is what Whewell calls the process that entrenches a hypothesis securely within the corpus of accepted scientific knowledge. It requires auxiliary assumptions that supplement the hypothesis in deriving empirical results to promote the simplicity and coherence of the resulting theoretical system; they must not appear arbitrary or be artificially contrived for the purpose of yielding those results. The difference between harmony and contrivance is to be measured by the ability of the resulting system to anticipate results *other than* those that the hypothesis was based on or intended to yield in the first place. A result whose derivation confirms a theory by thus attesting to the theory's naturalness and plausibility must differ from any results involved in the foundation or development of the theory.[7]

But what makes a result "other" than these? What principle of individuation is intended? While Whewell's own emphasis is on the independence of the theoretical motivation for supplementary assumptions from the results to be derived, the *problem* he introduces is to determine what qualifies further results as sufficiently different from those originally in view to establish the system's harmoniousness. He requires a criterion for the independence of a result whose successful prediction is to confirm the hypothesis from results whose prediction is to be expected whether or not the hypothesis is correct.

Some differences among results will obviously be incidental to the evaluation

---

6. A hypothesis that thus resists confirmation may yet be classified as "confirmable" on semantic grounds. But this simply means that logically possible situations instantiate it, and it is a mistake to assume that instances of a hypothesis are automatically confirmatory. That instantiation is *not* confirmation is clear if we consider empirical results that a theory predicts only in virtue of having assumed them in its formulation. Such results do not support the theory, for, as we saw at the end of chapter 1, no measure of truth is imputed to the theory in explaining how it manages to predict them successfully.

7. William Whewell, *The Philosophy of the Inductive Sciences* (London: J. W. Parker, 1840; reprint ed., New York: Johnson Reprint Corp., 1967) vol. 2, pp. 68–70.

of the hypothesis. Significant results are supposed to be repeatable, yet histori-cally, few, if any, experiments are repeated exactly. There are always intended variations, as well as those changes of materials and apparatus that make a new test a repetition rather than the original experiment itself. A criterion of indepen-dence that allows repetitions—however different physically from the original—to qualify as confirmatory is too weak. Any hypothesis should satisfy it. Simply to rule out repetitions as such—to trade on that intuitive notion—is likewise inade-quate, both because the notion is vague, and because genuinely different results might not be sufficiently different to carry epistemic significance. A more sophisti-cated analysis of independence is required.[8]

The consilience of inductions, as Whewell developed it, is not a successful method for warranting acceptance of hypotheses. The gap between consilience and truth looms, and an operational standard of consilience in terms of the deri-vation of additional results is elusive. Yet the methodological *direction* that Whe-well adumbrated might yet succeed, in the form of a deeper understanding of the independence of results and of exactly what such independence betokens. A number of philosophers have thought so, and it is to their more recent contribu-tions that I now turn.

Temporal Constraints

Imre Lakatos was probably the most influential philosopher to make the indepen-dence of predicted results a standard of epistemic warrant for theories. For La-katos, warrant attaches directly not to theories individually but to connected se-quences of them, called "research programs." The form of independence that enables an experimental result to confirm a research program he calls "novelty." And the importance of confirmations is to demonstrate that a research program is "progressing," rather than "degenerating." Progress is an epistemic notion for Lakatos, in that it provides rational justification for a scientist's allegiance to a program, for his decision to work within it and use it to explain and interpret experience. Whether such allegiance amounts to acceptance of the program's theoretical commitments, and whether, if it does not, such acceptance is also justifiable, are not consistently clear in Lakatos's writings. Either way, progress is epistemic because it provides rational grounds for theory choice, and Lakatos assumes that the historical record of chosen theories constitutes a growing body of knowledge.[9]

8. A good example is the history, from Michelson's work in the 1880s to the 1920s, of interferomet-ric experiments to measure the relative velocity of light. Such experiments differed in seeking to elicit a variation in velocity under conditions to which earlier experiments were insensitive. Presumably such differences do not matter to the question of epistemic support for relativity theory.

9. This does not mean that Lakatos regards it as *irrational* to prefer a degenerating program, or one not progressing as rapidly—producing novel results as frequently—as another. Degeneration could be temporary, and there is no point at which allegiance to a program is definitively irrational. More-over, Lakatos holds that the fortunes of a research program are as much a function of the aptitudes

Lakatos's use of the term 'novel' emphasizes that to attest to a research program's progressiveness, a result must be *new* to the program. The idea is that a novel result is one that only the latest theory in a program predicts. As the program develops, its basic ontological and metaphysical commitments are refined, and their applicability to scientific problems increases through the introduction of additional hypotheses, preferably hypotheses suggested by the program's originating assumptions. The addition of hypotheses marks the distinction between one theory of the program and the next. For this development to be progressive, a theory that results from the addition of hypotheses must make predictions not formerly yielded by the program. Should additional hypotheses merely serve to correct outstanding problems—to account for previous failed predictions or solve other difficulties internal to the program—without in turn yielding further predictions, then we do not get Whewellian consilience and the program degenerates.

Now, this does not so much address the issue of what *makes* a result novel as it does the issue of what of epistemic significance is *to be inferred* from its novelty. By focusing on what is inferable, Lakatos evades the problem of what results are independent or sufficiently different from results already predicted for their prediction to be new. He frequently writes as though this problem can just be left to intuition, or to the retroactive consensus of scientific judgment as to a research program's overall success.[10] His position is reminiscent of that of Adolf Grünbaum, who, having identified this problem, despairs of a general solution and proposes to deal with it only on a case-by-case basis.[11]

The standard interpretation of Lakatos on this point has it that novelty is a matter of *temporal* newness: Novel results are unknown and unanticipated before they are derived from a program, which thereby becomes progressive. In effect, novel results are *discovered* by working out the implications of the program.[12] And Lakatos does often speak this way. For example, he writes, "A new research pro-

---

and persistence of its practitioners as of the dispositions of nature. A scientist could justify holding to a relatively weak program on the grounds that its competitors are unfairly privileged. It is largely on the basis of these views that Lakatos is often held to have acquiesced in the relativism associated with Thomas Kuhn.

10. An important caveat must be attached to any discussion of Lakatos's views on rationality: Rationality, for him, is always a *retroactive* assessment that depends on how a research program plays out in the wake of the achievements on which its progressiveness depends. This is partly because Lakatos regards the relation of evidence to theory as complex and subject to reinterpretation. The progressiveness of a program is measurable in real time; the rationality of its victory over others requires hindsight.

11. Adolf Grünbaum, "The Bearing of Philosophy on the History of Science," *Science*, 143 (1964):1406–1412.

12. For example, Ian Hacking so interprets Lakatos, supposing the interpretation uncontroversial. The problem Lakatos addresses, Hacking thinks, is how to account for the importance scientists attach to predictions of previously unknown facts, when Mill, for example, has argued that it is only the ability to account for facts as such that matters, not when the facts are recognized. Hacking credits Lakatos with being the first "to make sense of the fact that prediction and not mere explanation can be so influential." See Hacking's review of Lakatos's *Philosophical Papers*, in *British Journal for the Philosophy of Science*, 30 (1979):381–402.

gram . . . may start by explaining 'old facts' in a novel way but may take a very long time before it is seen to produce 'genuinely novel facts,'" suggesting that (genuine) novelty requires temporal newness.[13] He also appears to use 'new' interchangeably with 'novel.'[14] The impression that this was his intended analysis of novelty was solidified by a proposal of Elie Zahar's to "modify" Lakatos's methodology, and Lakatos's subsequent "acceptance" of the modification.[15]

I doubt, however, that Lakatos *ever* intended temporal newness to be a necessary condition for novelty in scientific research programs. In first introducing novelty as a requirement for progress, he says only that a fact novel for a theory must be "improbable in the light of, or even forbidden by" the theory's predecessor. This requirement is surely satisfiable by facts long known ("old facts") and anomalous for the predecessor.[16] Add to this that he counts the precession of Mercury's perihelion as novel for general relativity and the Balmer series as novel for Bohr's theory of the hydrogen atom, although these results were established before those theories were introduced. Equally significant is Lakatos's endorsement of a broad conception of prediction that explicitly includes "postdiction" in his original analysis of progress in terms of predictive novelty.[17]

The importance of this point is not just exegetical. While the temporal order of theory and experiment may afford a convenient, workable criterion for de-

---

13. Imre Lakatos, "Falsification and the Methodology of Scientific Research Programs," in Imre Lakatos, *Philosophical Papers*, vol. 1 (Cambridge: Cambridge University Press, 1978); see p. 70.

14. For example, "Falsification and the Methodology of Scientific Research Programs," p. 34.

15. Elie Zahar, "Why did Einstein's Programme Supersede Lorentz's?," *British Journal for the Philosophy of Science*, 24 (1973):95–123. Zahar wants the Michelson-Morley result to count as a novel success for special relativity, although it antedates the theory by seventeen years. Lakatos agrees and deploys a modified account of novelty to explain why the Copernican program supplanted Ptolemaic astronomy. Facts are now to count as novel *for a program* if only the program was not based on them. As Hacking tells it, "Facts need not be strictly speaking novel [because] they may have been known before," yet they are to "count as 'novel' if they were not considered as part of the original inspiration of the program." Hacking goes on to lament Lakatos's acquiescence in a looser, psychologized, program-relative account of novelty.

16. "Falsification and the Methodology of Scientific Research Programs," p. 32. Michael Gardner, exemplifying the standard interpretation, claims that Lakatos's requirement that a novel result be "improbable" or "forbidden" precludes previously known results. Gardner does not explain how this preclusion is supposed to work, but evidently he means that once a result is known it cannot be considered improbable. Probability, however, is conditional, and what Lakatos conditionalizes a novel result's probability on is nothing so loose as past observations or information, but the *preceding theory* in a research program. Gardner elects to quote from a footnote in which Lakatos speaks of "previous knowledge," but the more careful and detailed text to which that very footnote directs the reader for clarification refers explicitly to preceding theory. It is unclear why Gardner assumes that the footnote takes priority. The reason that the footnote speaks more generally of "knowledge" is only that the example in the footnote is of an empirical generalization rather than a theory. What Lakatos thinks precludes the novelty of the next instance of an empirical generalization is not the failure of the next instance to be new, but the fact that previous instances provide reason to expect it. Whether he is right in so thinking is a question I will consider. See Gardner, "Predicting Novel Facts," *British Journal for the Philosophy of Science*, 33 (1982):1–15; esp. p. 2.

17. "Falsification and the Methodology of Scientific Research Programs," p. 32.

limiting a theory's prospective support, it is surely unintuitive to make one's episte-mology depend on what could, after all, be simple happenstance. Should a cer-tain experiment happen to have been delayed—through the vagaries of funding, perhaps—then a theory without support becomes supported. The temporal crite-rion was historically rejected by such philosophers as Mill and John Maynard Keynes, essentially because the timing of an experiment might have nothing to do with that of a theory's development.

Intuitively, many classic experimental results have a preanalytic claim to nov-elty with respect to later theories credited with being the first to explain them successfully. And intuitively, 'novel' does not mean simply "new"; when scientists mean "new," they say it. 'Novel' is used, rather, for newly explained or predicted results that are different, unusual, or anomalous with respect to background the-ory, and that may or may not be newly disclosed in the course of predicting them. The Michelson-Morley experiment to measure the relative motion of the earth and the electromagnetic ether, the precession of the perihelion of Mercury, and Galileo's law of falling bodies are probably examples; they have been regarded, correctly or not, as novel, independently of any particular attempt to explicate the concept of novelty in general. Indeed, it is often considered a constraint, or condi-tion of adequacy, on the philosophical analysis of novelty that certain results, such as these, qualify. Most important for my purposes is that the point of ensur-ing novel status for these results has been to ensure that their probative character, with respect to the theories for which they are novel, be respected. That is, the connection between the novelty of a result and the evidential bearing of the result on the theory for which it is novel is just assumed.

Zahar, for example, is intent on making the Michelson result novel despite its temporal antecedence, *so that* it can be grounds for replacing ether-based electrodynamics with special relativity. And Lakatos wants to allow various facts about planetary motion known since antiquity to count as novel for Copernican theory, so that his methodology of research programs can justify the success of Copernicanism. Theories often receive strong support from their ability to predict previously recognized results, and if Lakatos were imposing a temporal constraint on novelty, he would be contesting historical judgments of evidential warrant that underwrite his own analysis of theory choice in terms of the methodology of research programs.

It seems that two desiderata for theories coalesce under the concept of novelty, and Lakatos has not consistently marked their distinctiveness. On the one hand, there is an interest in empirical discovery. We want to increase our knowledge of the natural world, and good theories function as tools toward that end. It is one measure of a theory's value that by applying it, we reveal and identify physical phenomena not previously recognized. This desideratum leads us to promote the-ories that produce results novel in the sense of temporal newness. On the other hand, there is an interest in learning why physical phenomena exhibit the pat-terns they do, in explaining the natural world. Theories propose such explana-tions, and we are interested in evaluating them. Successful prediction counts to-ward a favorable evaluation of a theory if the result predicted is not one that we have reason, independent of the theory, to expect. For then the success of the prediction must be credited to the theory that produces it, if to anything. Thus, the desideratum of explaining natural phenomena leads us to promote theories that make successful, novel predictions, but in a sense of 'novel' that pertains to independence rather than temporal newness. What seems to matter to evaluation, in contrast to discovery, is that predicted results be *new to the theory*, so that the theory does not depend on their being available antecedently; it does not matter that they be new to our knowledge or recognition.

Bayesian Confirmation

The plausibility of a temporal understanding of novelty is often discussed in connection with Bayesian confirmation theory. Roughly, this is the theory that evidence confirms a hypothesis by increasing its probability in accordance with Bayes's theorem of probability. In general, we express the probability of a proposition p, estimated on the assumption that a proposition q is true, by "pr(p/q)," where, according to the definition of conditional probability,

$$\text{pr}(p/q) = \text{pr}(p\&q) \div \text{pr}(q).$$

Let h be a hypothesis; e be the evidence that is to confirm h; and b be background information available antecedently to the introduction of h, including other hypotheses, theories, and evidence. The probability pr(h/e) is the probability that h is true after e is taken into account. In the "personalist" interpretation of probability favored by (most) Bayesians, this means the degree of belief or confidence that a (ideally rational) cognitive agent who knows of e has in h. The probability pr(h), or pr(h/b) in conditional form, represents the probability of h independent of e, or prior to the availability or consideration of e, with only the general background knowledge b taken into account. Bayes's theorem in its simplest form may then be written as

$$\text{pr}(h/e\&b) = (\text{pr}(h/b) \times \text{pr}(e/h\&b)) \div \text{pr}(e/b).^{[18]}$$

The theory is that e confirms h if pr($h/e\&b$) > pr(h/b); that is, if the addition of e to the antecedently available information increases the probability of h. pr($h/e\&b$) − pr(h/b) is then a natural measure of e's confirmation of h. According to the theorem, confirmation requires that

$$\text{pr}(e/h\&b) \div \text{pr}(e/b) < 1.$$

18. To obtain this form of the theorem, use the definition of conditional probability to write the probability of the conjunction of h and e twice, reversing the order of the conjuncts. Then solve for pr($h/e\&b$).

Since *e* is supposed to be evidence for *h*, it is natural to assume that *h* predicts *e*, so that *e* will be deducible from *h*, together with collateral facts contained in *b*.[19] Then pr(*e*/*h*&*b*) = 1, and confirmation depends on having pr(*e*/*b*) < 1.

However, if *e* is known prior to the introduction of *h*, then *e* will be part of *b*, so that pr(*e*/*b*) = 1. Therefore, if no temporal constraint is imposed on novelty, *e* can be a successful, novel prediction of *h* without confirming *h* *at all*. Bayes's theorem gives us reason to require not only that a novel result be temporally new—discovered through working out the implications of the theory or hypothesis that it is to test—but, more generally, that it be a result that, as Lakatos would have it, previous knowledge renders unlikely. The lower pr(*e*/*b*) is, the greater is *e*'s confirmation of *h*.

The only evident way to avoid this temporal restriction is to interpret *b* as devoid of *e* even where *e* is part of the background knowledge. We would have to construct some artificial *b'* = *b*−{*e*} for the probability of *h* to be conditionalized on, and then interpret pr(*h*/*e*&*b'*) as the degree of confidence that would hypothetically have been placed in *h* were *e* to have just been learned. This is not really satisfactory in the evaluation of actual historical cases, in which the infusion of *e* into the background knowledge is such as to render *b'*, so defined, scarcely intelligible. Theorizing may have been conditioned by the requirement of consistency with *e*, or by the need to explain *e*, so that one really cannot say what the background knowledge would be with *e* left out of account.[20]

As we have found reason to fault a temporal account of novelty, while upholding the significance of novelty for confirmation, we must reject the adequacy of Bayesian confirmation theory as an analysis of confirmation (though no one, of course, rejects Bayes's theorem itself as a mathematical relation among probabilities). More generally, I will reject reliance on the use of probability as an approach to understanding the relation of confirmation between theory and evi-

---

19. I ignore, for simplicity, the case of statistical hypotheses.

20. The seminal discussion of the problem that an atemporal view of novelty poses for a Bayesian theory of confirmation—the problem of "old evidence"—is Clark Glymour's in *Theory and Evidence* (Princeton: Princeton University Press, 1980), chap. 3. Howson and Urbach, in *Scientific Reasoning* (La Salle, Ill.: Open Court, 1989), chap. 11, argue that if old evidence really posed the problem for Bayesianism that Glymour claims it does, then so would *any* evidence. Evidence that becomes available after a hypothesis is proposed nevertheless, once learned, enters the background-knowledge base. And if we are allowed to calculate the probability of a hypothesis conditional not on the total knowledge situation but on knowledge from which newly learned evidence is hypothetically deleted, then we should, equally, be allowed to conditionalize the hypothesis on knowledge from which old evidence is hypothetically deleted.

But this claim ignores the growing effect of evidence on the knowledge situation over time. It is one thing to be able to characterize the knowledge base on which a hypothesis is conditionalized, with a newly discovered result hypothetically (or counterfactually) suppressed, and quite another to be able to do this for a result that has already influenced the course of science. Moreover, Howson and Urbach ignore the option, unavailable in the case of old evidence, of calculating the probability of a hypothesis conditional on its predictions before those predictions have been tested. If we make this calculation and stick with it after the prediction is borne out by actual evidence, we obtain a measure of the confirmation afforded by the evidence that is simply unavailable in the case of old evidence.

dence. The development and evaluation of that approach constitute an enormous topic. I will defend my rejection of it with only a few points that apply specifically to the use of Bayes's theorem.

One is that the confirmation of a hypothesis on the basis of any evidence will be greater, according to Bayesian theory, the more probable the hypothesis is in light of background information. Hypotheses that are unprecedented or radical in nature are correspondingly harder to support. Although the *impact* of e on h, measurable, for example, by the difference between pr(h/b&e) and pr(h/b), need not be reduced for a radical hypothesis, the role of b in the theorem—the fact that pr(h/b&e) is directly proportional to pr(h/b)—establishes a preference for perpetuating entrenched theoretical approaches. While there is some intuitive appeal to the idea that the evidential burden faced by a hypothesis strikingly different from those already found plausible is greater, the evident conservatism that the theorem, read as the measure of evidential support, expresses may give one pause. The theorem tells us that even a hypothesis fully in accord with and explanatory of all known empirical phenomena is automatically less acceptable and harder to support, just because it is different.

If, for example, a wave theory of light is in vogue, then a new theory that treats light as particulate is automatically discredited, however well it compares to the best extant wave theory on the basis of available evidence. Historically, some very important and successful hypotheses—Einstein's explanation of the photoelectric effect in terms of a quantum structure for light, for example—have been introduced in just such circumstances. And correspondingly, there is a tradition in empiricist philosophy—well represented by Popper—of *favoring* not only novel evidence, but hypotheses and theories that appear novel, with respect to established ideas, because of their audacity and originality, provided that they pass muster empirically.

A different sort of problem is raised by the Bayesian's reliance on the personalist interpretation of probability to supply the "prior" probabilities, pr(e/b) and pr(h/b), needed to apply the theorem. The fact is that there simply is no general, objective method of numerically estimating these values. The idea that one can attach numbers to these expressions in any meaningful way is fanciful.[21] Personalism acknowledges this, and proposes, in the absence of objective values, to use subjective ones, constrained only by consistency with the axioms of probability.

This constraint takes the form of an appeal to ideally rational cognitive agents, and it is ironic that while individual probability assignments are held to be subjective and hence unrestricted, all cognizers are expected to be immediate and unquestioning Bayesian conditionalizers. The idea seems to be that epistemic rationality has nothing to do with the content of one's beliefs, and everything to do

---

21. It is reminiscent, in *Star Trek*, of Mr. Spock's habit of quoting enormous odds to indicate the hopelessness of some course of action that Captain Kirk is contemplating. There is no basis whatever for such calculations. I realize that it's the twenty-third century, but these probabilities are right up there with molecular transporters, universal translators, and warp drive.

with how beliefs are processed—how they are modified or replaced as one's epistemic situation changes. There is *nothing* that it is irrational to believe, whatever one's evidence, provided only that all combinations of one's beliefs conform to the rules of probability. Why the scope of rationality should be restricted in this way is unexplained.[22]

Thus, personalism allows radically divergent probability values to be assumed without any relevant difference to explain them. It is a mathematical fact that such disparities eventually dissipate under the weight of increasing new evidence, provided that probabilities are adjusted by Bayesian conditionalization. But "eventually" can be a long time. This result hardly compensates for the large, unexplainable differences in confirmational import that any given body of evidence is allowed to have. The Bayesian can only attribute these differences to arbitrary differences in prior probabilities. Everyone is supposed to operate with the same (Bayesian) methodology, but no one is, in the least, constrained as to the confirmational significance attached to any given piece of evidence, nor to initial estimates of plausibility. Bayesianism purports to be a quantitative theory of confirmation, and, in this respect, to improve on earlier theories. But personalism renders the promise of numerical measures of confirmation hollow.

Even worse, Bayesianism's reliance on personalism invites an evasion of serious methodological issues. If different scientists respond differently to the same evidence, one treating it as highly significant, while the other dismisses it, it may well be that there is a deeper disagreement at the methodological level as to the standards by which theories are appropriately judged. For example, one might insist that confirmatory evidence be novel, while the other thinks this immaterial. Rather than addressing such a question, the Bayesian will characteristically attribute the difference in evidential weight to a difference in prior probabilities, resorting to a subjective analysis on which the apparent disagreement is unreal. Bayesianism becomes a stratagem for pretending that methodological differences do not require rational adjudication. It is not recognized, or not admitted, that the confirmation of substantive theoretical beliefs inherits the subjectivity and irrationality on which this stratagem relies.

This concern highlights a general limitation of the Bayesian approach. The construal of confirmation in terms of probability, via Bayes's theorem, leaves no

22. The alternatives may be compared via an analogy between foundationalist and coherentist epistemological theories, on the one hand, and democratic and totalitarian legal theories, on the other. In a democracy, everything is legally permitted that is not explicitly proscribed by statute (or by the interpretation of statute by case law); the burden is on the state to show that an action is impermissible. In a totalitarian system, everything is prohibited except what is officially approved, in that the state, in its sole discretion, can punish any action without prior legal sanction. The burden is on the individual to show that an action is permissible. Foundationalism, by analogy to totalitarianism, requires beliefs to be justified. Coherentism, by analogy to democracy, permits all beliefs, save those that can be refuted. Foundationalism burdens the believer; coherentism, his critic. Typically, then, foundationalism inquires into the warrant for beliefs, while coherentism inquires into the warrant for *changes* in beliefs.

What is good for law is not necessarily good for epistemology.

room for the methodological dimension of evaluation of theories. It presumes that all that ultimately matters to the epistemic standing of a theory are the content of the theory, the relevant background theories, evidence, and related auxiliary information. How the theory was developed, and whether the methods used have independently been found fruitful or successful in other cases or have been historically discredited, are deemed irrelevant. Also irrelevant is whether or not the theory was designed to yield the evidence, whether it was so constructed as to guarantee, in advance, that the evidence would be predicted. In compliance with Hans Reichenbach's admonition that the process of theorizing—the context of discovery or invention—be dismissed from the epistemic enterprise as mere psychology, Bayesianism assumes that a theory's epistemic credentials are unaffected by its provenance.

The basic Bayesian measure of epistemic support for a theory T by observable condition O, $S(T/O)$, equates it to the difference between T's posterior probability, conditional on O, and T's prior probability:

$$S(T/O) = \mathrm{pr}(T/O) - \mathrm{pr}(T).$$

Using Bayes's theorem to rewrite $\mathrm{pr}(T/O)$ delivers

$$S(T/O) = (\mathrm{pr}(T) \div \mathrm{pr}(O)) \times (\mathrm{pr}(O/T) - \mathrm{pr}(O))\,[23]$$

so that support is proportional to the difference between O's probability conditional on T and its prior probability. But this factor can be made arbitrarily large under conditions that would not warrant believing the theory, under conditions in which O is not really evidence at all (which is why I use 'O' here rather than 'e'). Suppose that the theorist conjectures an unknown O and builds T on it, so as to guarantee that O is deducible from T. T need not provide the only, nor even a very good, explanation of O; yet should the conjecture turn out correct, T is pronounced strongly supported.

The probative significance of empirical evidence for a theory depends on its independence from the theory's provenance in ways that the Bayesian measure cannot capture. We have seen that the explanation of a theory's predictive success—in particular, whether this explanation need impute truth to the theory—depends on the involvement of the predicted phenomena in the theory's development. If novel status is to reflect such independence, the context of discovery must provide a crucial determinant of epistemic standing.[24]

23. I here pursue a suggestion of Graham Oddie in comments delivered at the American Philosophical Association's 1995 Pacific Division meetings in San Francisco.

24. For interpretation of Reichenbach's distinction between the contexts of discovery and justification, and defense of the epistemic relevance of discovery, see J. Leplin, "The Bearing of Discovery on Justification," *Canadian Journal of Philosophy*, 17 (1987):805–814.

Contemporary Analyses of Independence

At this point dangers of imprecision loom in the terms available to explicate novelty. For example, the expression "new to the theory," by which I have stipulated that a novel result be one whose antecedent availability a theory not depend on, suggests that a novel result may have already been predicted or explained by *other* theories. But in that case we *do* have reason independent of the theory under evaluation to expect the result, and its successful prediction need not be credited to the theory. What "new to the theory" is meant to suggest is, rather, that the result have had no role in developing or motivating the theory, no role in the theory earlier than its confirmatory role, once the theory is ready for empirical evaluation. However, the notion of a "role" in development or motivation is also unacceptably vague, as some such roles may be innocuous with respect to a result's probative significance. Does the precession of Mercury's perihelion fail to support general relativity because Einstein hoped or intended that his theory would account for it? What does it matter to the epistemic standing of a theory what its inventor hoped or intended? And if the psychological dimension of theory construction is allowed to preclude probative significance, how can we exempt judgments of the epistemic standing of theories from the notorious unreliability of estimations of psychological conditions?

The term 'independence' is equally problematic. A result predicted by a theory will not be independent of the theory in the logical sense, for a deductive relationship will obtain. The point of requiring independence is to ensure that a result carrying the epistemic significance associated with novelty will not be one to which the theory owes its provenance or rationale. For if this sort of dependence obtains, then the fact that the result is predicted by the theory is built in from the beginning and requires no further explanation. There may be reason, independent of the theory, to expect the result—reason in the form of whatever supplied the result in the first place—and this sort of independence we do not want. Further, a result on which the theory does not depend in the intended sense will not only be independent *of the theory* in this sense, but will also, thereby, be independent *of other results* on which the theory *does* depend, in this sense. Thus, we are vulnerable to a potentially confusing slide from talk of the independence of a result from a theory to talk of independence as a relation among results. A result's independence from other results on which a theory depends is necessary but not sufficient for independence from the theory. The latter independence is the fundamental one, and an adequate account of it will provide for independence among results.

Unfortunately, influential extant analyses of novelty typically run afoul of these pitfalls, and of others that these generate. The analysis by which Zahar proposes to admit past results construes novelty as independence from the *problems* that a hypothesis is intended to address. This move immediately relativizes novelty to a particular hypothesis or theory; a result may be novel with respect to one theory but not another. This feature is already implicit in Lakatos, via the conditionalization of probability. He requires that the probability of a novel result be low, measured on the condition that preceding theories are correct. This much relativization is both innocuous and unavoidable; if novelty is to be an evidential property, what a novel result is evidence for must be specifiable. But relativization to theories suggests that a result predicted by several theories may be novel for just one of them. In this case, the explanation of the prediction's success need not impute truth to the theory for which it is novel. An analysis that preserves the probative import of novelty must proscribe this situation.

A more pressing problem is that Zahar's proposal opens a psychological dimension to novelty. What problems a hypothesis is intended to address depends on the knowledge, interests, and expectations of the theorist. Those who think Lakatos held a temporal view of novelty credit him with sensitivity to the dangers of allowing novelty to depend on such variables. If the theorist happened not to intend to account for a result that, under changes of knowledge or interest, he would have intended to account for, the result acquires an evidential importance that it would have lacked but for those changes. So what theories are supported and how well they are supported are potentially hostage to idiosyncrasies of the theorist. This seems an irremediable liability of understanding independence in terms of problems.

Perhaps in an (unconscious?) effort to avoid this outcome, Zahar makes the *use* of a result in constructing a theory his operative criterion for novelty, despite his talk of problems. He defends the novelty of results, whether antecedently known to the theorist or not, on the grounds that they were not used. This move does not, however, obviate the role of the theorist. Perhaps another theorist would not have used the result; would it then have been novel? Novelty is again relativized to the theorist as well as to the theory.

It is not clear that talk of "use" can avoid a psychological or otherwise extra-epistemic dimension to novelty. To what extent a scientist relies on empirical results in fashioning a theory could be a matter of aptitude and inclination. The availability of empirical results could depend on funding opportunities, or be otherwise related to social priorities. Perhaps a scientist believes that empirical data are necessary to fix parameters in a theory, but in fact a theoretical argument unknown to him would suffice. Had he used such an argument, results on which the theory relied would instead have achieved the status of confirmation. Perhaps two scientists approach a problem differently, one relying on results unknown to or uninfluential for the other, who develops a theoretical argument leading to the same solution. Can the epistemic standing of a theory be allowed to vary with such differences?

What is to constitute use? Is a result used if thinking about it helped the theorist formulate his ideas; that is, is a purely heuristic role to constitute use for purposes of the analysis? Is a result used if cited by way of illustration or example, or must it have been impossible to have constructed the theory without knowledge of the result? Should a diagnosis of use depend on whether the result is cited in written reports of the theory's origination? Or is it enough that references to the result appear in presentations of the completed theory, regardless of the result's actual role in the theorist's own work? The possibilities vary in the infusion of psychological components. Even an operational approach using written citations to circumvent the psychological dimension of use does not prevent a theory's epistemic standing from varying with incidental, contingent features of its genesis.

Be that as it may, it is not clearly desirable, even if possible, to make just any intrusion of psychology defeat novel status. All sorts of reflections could be psychologically influential without thereby forfeiting their justificatory significance. If thinking about the irrelevance of mass in free fall led Newton to relate weight to orbital motion, does the law of gravity gain no support from its ability to yield Galileo's law? Perhaps so; perhaps Galileo's law supports Newton's law of gravity only because Newton's law *corrects* Galileo's, in addition to yielding it as an approximation. But a psychologically influential result might be too incidental to the theory it influences to be denied a justificatory role. Before developing general relativity, Einstein believed that the precession of Mercury was a relativistic effect, and he used the Mercury case as a criterion of adequacy for theories he considered.[25] Yet he cited Mercury's precession as confirmation of the final theory, and the scientific community supported that assessment. Can the precession of Mercury not be novel for general relativity, despite being so used?

One reason for objecting to a criterion that depends on psychological features of the individual scientist is that these may be inaccessible. It becomes a daunting historical problem to determine that a result is novel, if its novelty would be defeated by facts that are unlikely to be widely recognized or included in published documents, facts that may not be recognized even by the scientist himself. This concern may be assuaged by appealing to the state of knowledge or the practice of the scientific community, or a relevant portion of it, rather than to the individual scientist. That a result is not among those that a new theory in a given field would be expected to address, or among those that are generally known to the scientific community and available for use, are more reliably attestable conditions than what an individual scientist knows or intends. Perhaps *sociologizing* novelty is less cumbersome than psychologizing it. Historical case studies in theory construction can show, quite reliably, what was expected of *any* new theory that purported to accomplish what the ultimately successful theory in a given field accomplished; they can document the known results that scientists seeking a new theory had to work from. At least for major physical theories of the nineteenth and twentieth centuries, it is unnecessary to speculate as to what was known and what were the standards of evaluation in terms of predictive or explanatory tasks.

However, such a community-relative criterion of novelty promises neither a necessary nor a sufficient condition for justificatory status. Michael Gardner recognizes the insufficiency, although the only reason he gives for it is his presumption that a result previously known, whether by the whole community or just the

---

25. Einstein wrote to Conrad Habicht, on December 24, 1907, of his determination to solve the problem of Mercury. At one point (1913), Einstein seemed willing to accept a theory that failed to solve the problem, because he feared that this would be the cost of achieving general covariance. See Roberto Torretti, "Einstein's Luckiest Thought," in J. Leplin, ed., *The Creation of Ideas in Physics* (Dordrecht, Neth.: Kluwer Academic Publishers, 1995).

scientist whose theory predicts it, cannot be novel. I have suggested that this assumption is dubious, and will offer a more important reason independent of it. A result that is not known to anyone, and, thus, no part of the standards for evaluating prospective theories, might nevertheless be a (unrecognized) consequence of an already developed theory that is incompatible with the new theory for which the result's novelty is at issue. That is, a result that satisfies a community-relative criterion, and is thereby supposed to come out novel for a new theory from which it is first derived, might already be deducible from a competing theory in the field. In that event, there is antecedent reason, albeit unappreciated, to expect the result independently of the theory for which it is thought to be novel. And, as previously urged, this circumstance undermines the probative status that novelty is supposed to ensure.

Gardner supposes that communal knowledge and standards do supply at least a necessary condition. And this does not square with the preanalytic claim to novelty of the classic results I have mentioned. To revert to the least controversial case, although the precession of Mercury's perihelion had been an object of investigation and attempted explanation for over half a century, its derivation from Einstein's 1915 (November 11) version of relativity theory was counted an impressive success, and convinced Einstein himself that he finally had the correct field equations.[26] A challenge to the novelty of this result requires a principled, theoretical basis; it is not to be dismissed through a facile identification of novelty with "newness." "Plainly," writes Gardner, "no one is going to take seriously as a 'prediction of a novel fact' a proposition which the investigators in a given field were taking for granted and trying to explain."[27] This is far from plain. It is ironic to disqualify results that a theory is expected to account for as a criterion of its success, when the importance of novelty is to capture the features that make a correct prediction especially probative for a theory.

The historical problem of making novelty too recondite an attribute to be reliably assessed suggests the substitution of communal for personal standards and knowledge, but the substitution does not work. Gardner's response is to finesse the historical problem. He makes novelty entirely a matter of individual ignorance, proposing to regard as novel for a given theory only results unknown to the theorist at the time he constructs the theory. It may be difficult to tell what these are, but at least it will not generally be as difficult as determining what results are influential or intended as outcomes. There is no perfect solution, thinks Gardner, and the advantage of this solution is to err on the side of rigor. Unknown results cannot very well be influential or intended.

For this view of the matter to persuade, we must be prepared to allow, with Gardner, that if a result somehow *were* certifiable as uninvolved in a theory's development, this would be enough to establish its novelty. But as I have noted, the probative weight that novelty is to carry entails further constraints, and is not

26. See John Norton, "Eliminative Induction as a Method of Discovery: How Einstein Discovered General Relativity," in J. Leplin, *The Creation of Ideas in Physics*.

27. Gardner, "Predicting Novel Facts," pp. 9–10.

necessarily defeated by all forms of involvement. A result unknown to a theorist and, hence, uninvolved in his theory could be known to others, could even be regarded as successfully explained and predicted by a competing theory. A theory's ability to account for a result can redound to its credit, even if—indeed, partly because—there was antecedent reason to regard the result as one that an acceptable theory ought to explain. A result that influenced or motivated the theorist might yet be novel for his theory if that influence was so incidental to the theory that the course of the theory's development would have been unaffected had the result been unknown.

Gardner's desire to settle on a simple and historically manageable criterion only partly explains his failure to be impressed by such concerns. More telling is that his basic approach to the problem of novelty does not respect the connection of novelty with epistemic significance, on which I have been insisting. This failure is illusive, because Gardner identifies the support that novelty provides as the motivation for his investigation. Once motivated, however, his investigation proceeds in complete independence of novelty's epistemic significance. Only when his final, ignorance criterion is in place does Gardner ask how novelty accomplishes its justificatory role, concluding, ironically if unsurprisingly, that it doesn't.[28] The near universal practice of according novelty special probative force, which Gardner acknowledges and sets out to rationalize, he finds to be devoid of justification. By my lights, this outcome undermines not the importance of novelty, but the adequacy of Gardner's analysis of it.

Whether an epistemically adequate analysis inevitably runs afoul of historical tractability remains to be seen, for no such analysis is yet in view. But I wish to point out that novelty's epistemic role has a conceptual basis. The very point of *calling* a result "novel", rather than simply "new" or "unexpected," is to register its epistemic importance. Gardner concedes as much in motivating his inquiry, but does not recognize the constraints this concession places on admissible analyses. An analysis that does not support valuing a successful novel prediction over just any consequence of a theory fails to register a significant distinction in the language.

Of course, an epistemic point is only *part* of the import of diagnosing novelty, as Lakatos clearly saw. Lakatos originally stressed the *anomalous* character of novelty. A novel result was to be one unforeseen by preceding theories or, better, rendered *unlikely* by preceding theories. Alan Musgrave, pursuing this suggestion, proposes to regard any result a theory correctly predicts as novel confirmation if only it is not predicted by the best extant alternative theory.[29] On this view, a result's novelty relates it not only to the theory that predicts it but also to other theories replaced by, or in competition with, this theory. This addition, ignored by Gardner, only serves to underscore the epistemic constraints on novelty.

28. More precisely, Gardner says he is unable to find any basis in "more general principles of reasoning" for an effect on the evidential weight of a result of the time at which it becomes known.

29. Alan Musgrave, "Logical versus Historical Theories of Confirmation," *British Journal for the Philosophy of Science*, 25 (1974):1–23.

For to the extent that alternative theories *do* prepare for, or anticipate, a result, they undermine the credit that the result brings to the particular theory for which it is supposed to be novel. Thus, the effect of Lakatos's original insistence on the anomalousness of novelty is to protect a novel result's justificatory impact.

A further disadvantage of Gardner's proposal is that it relativizes novelty *too far*. Novelty's epistemic role requires its relativization to theories. But like a criterion based on problems, a knowledge-based criterion risks introducing a dependence on reference to the individuals who constructed the theory, via their knowledge of particular results. I have objected to allowing a theory's support to vary with the identity of its inventor. And if novelty is to count epistemically, then Gardner's criterion would have us regard a theory as better supported the greater the ignorance of its inventor. Perhaps *the theorist* is to be more highly regarded the greater the disadvantages under which he labored, but what does this have to do with the importance of the theory? The problem is that we judge both the theory and the theorist, and these judgments are distinct; they answer to different standards and are apt to diverge. If August Kekulé discovered the structure of benzene rings in a dream, as is the lore, the stature of his theory does not redound to his personal credit to the extent that it would if he had worked out the structure in some intellectually estimable way.

Allocations of credit are disputatious. Heinrich A. Lorentz and Henri Poincaré were, at one time, held to be the true originators of relativity theory. One reason this matters is that the theory *itself* is supposed to be an important advance, and this judgment is, in the nature of the case, independent of how we allocate credit. While it might have been *more of an achievement* for one person to have come up with a theory than another—because of their respective backgrounds or resources—this difference does not reflect on the merits of the theory.

The theorist's hopes, expectations, knowledge, intentions, or whatever, do not seem to relate to the epistemic standing of his theory in a way that can sustain a pivotal role for them in an analysis of novelty. I have argued that a criterion of novelty based on use raises the same problem. John Worrall disagrees. In many cases, he contends, it can be objectively determined whether or not a result was used in constructing a theory without relying on facts specific to the theorist.[30] Worrall does not, however, offer use as a criterion of novelty. He assumes that 'novel' means "new," and proposes to allow results that are not novel but were not used to count in a theory's assessment. But if the theory-relativity of novelty is recognized, then there is no need for this complication. Temporally new or not, a result may be *new to the theory*, thereby qualifying as novel in a nontemporal, epistemic sense of 'novelty', without prejudice to Worrall's proposal.

The difficulty with Worrall's proposal is that even if it were made clear, in general terms, what counts as use of a result by a theory, it would not be possible to identify the category of supportive results with the category of results

---

30. John Worrall, "Scientific Discovery and Theory Confirmation," in J. C. Pitt, ed., *Change and Progress in Modern Science* (Dordrecht: D. Reidel, 1985).

successfully predicted but not used. This identification is both too exclusive and too inclusive; both too strong and too weak.

It excludes results that although used, need not have been. Suppose a theory that used a certain result could have been developed just as well without it, so that the use of the result is incidental to the theory. If it is nothing more than a historical accident that the result was used, then the intuition that militates against the temporal criterion of novelty also militates against precluding a justificatory role for the result.

Suppose that although not used, a result is similar in kind to results that were used. Suppose the similarity is such that the unused result might have served as well as those used, the substitution having no effect on the theory. For example, a result may be used by being generalized, its generalization being what is then important to the development of the theory. Other instances of the generalization, had they been available, could have served a role in the theory's construction identical to that of the result actually used. If it is a historical accident which results a scientist used, it is unintuitive to treat them differently for epistemic purposes. A result not used would seem to be disqualified by its interchangeability with results disqualified because they were used. Alternatively, we might decide that a used result is disqualified unfairly if relevantly similar to results that qualify because they were not used. It is questionable how powerful or helpful intuitions are at this point, but they carry far enough to fault a criterion of support expressed simply in terms of use. The fact that a particular result was used, or that it was not used, may be too incidental to affect a theory's epistemic standing.

Other concerns have been raised against a use criterion by Deborah Mayo.[31] She argues that a use criterion is insufficient because results not used might have been selected in a manner that is biased in favor of the theory, other results in conflict with the theory having been dismissed illegitimately. Or an interpretation favorable to the theory might have been enforced on the results. Or, though not used, they might be too weak to test the theory because the theory is not needed to deduce them. She argues that a use criterion is not necessary, because a result that was used might be strong enough to imply the theory on its own.

These are heavy failings of a use criterion, especially if the results it passes are supposed to warrant inference to the hypothesis they test. A criterion could establish a weaker relation of evidential support that does not warrant inference. Acceptance of a hypothesis depends on factors other than its relation to any given test—on the outcomes of other tests and on the availability of rival hypotheses passing the same test, for example. It is not quite clear which interpretation Mayo intends; there are passages in her text supporting each interpretation. She advo-

---

31. Deborah Mayo, "Novel Evidence and Severe Tests," *Philosophy of Science*, 58 (1991):523–553. My discussion of Mayo's position is directed at this article. Mayo's book, *Error and the Growth of Experimental Knowledge* (Chicago: University of Chicago Press, 1996), appeared after this book was in press. I thank Professor Mayo for attempting, in personal correspondence, to help me understand her position.

cates "severity" over a use criterion of novelty as the condition of evidential support, and she seeks a criterion for a result to be a "genuine" or "severe" test of a hypothesis. But even on the weaker interpretation, such support as accrues to a hypothesis from a result not used to construct it may be offset by the availability of rival hypotheses capable of predicting the same result.

Worrall does not address these concerns about the adequacy of a use criterion, but he offers something that responds to them in a general way. He proposes to block any infusion of subjectivity or extraepistemic influence into theory appraisal by making the relation of evidential support *tertiary*, rather than binary. He insists that the question of whether a result e supports a theory T is incomplete without reference to T's provenance, p. Then the triple (e,T,p) may exhibit the relation of support, S, written "$S(e,T,p)$," where p is free of e, while the triple (e,T,p') fails to exhibit this relation, hence $\neg S(e,T,p')$, because p' uses e.[32] Then the question whether e supports T has no definitive answer because it is incomplete; the sentence "e supports T" does not carry truth value. Deploying this device, we would say that if e was used but need not have been, then there is an alternative provenance available on which the evidential relation does not hold. Similarly, if e was not used but could have been, then the relation holds for T's actual provenance but not for another of T's possible provenances.

I hope that spelling out the proposal to make support tertiary suffices to reveal its inadequacy. The proposal does not eliminate subjectivity; it merely incorporates it into p. If there was a problem of subjectivity in the first place, it will remain on the tertiary reading. Worrall wants to deny that subjectivity is a problem. He thinks that it is not the identity of the scientist qua individual that matters to whether a result is confirmatory, but only the heuristic or method he employs that matters. The only argument he gives, however, is to declare confirmation to be a tertiary relation. He does not show how to omit facts about the individual from an adequate description of the course of theorizing that the individual follows.

This objection is surmountable, but there is a more important criticism. Our concern was that a result actually used might nevertheless support the theory, or that one not used would nevertheless fail to support the theory. If these are legitimate concerns, then *support cannot be tertiary*. For their very formulation requires a binary reading. They require that the question whether e supports T be capable of being raised, if not answered, independently of determining whether or not e was used in constructing T. They treat the fact as to whether or not e was used as a reason for or against imputing truth to the claim of support, rather than as a constituent of the claim to which truth value is imputed. Nor is this merely a formal difference. On the tertiary reading, we cannot meaningfully entertain the question of support prior to specification of p. We cannot meaningfully pose, as a conceptual issue, whether a result that supports a theory might not have supported it, for whether or not a relation of support holds is decided in a purely formal way by checking whether the constituent p in that relation includes e.

---

32. John Worrall, "Scientific Discovery and Theory Confirmation."

We have seen that there is a straightforward rationale for regarding novelty as at least binary, relating a result to a theory. A result that is novel for one theory need not be novel for another. The rationale is simply that the concept of novelty has epistemic significance. Its use must therefore include reference to that of which knowledge is at stake, as the relational character of support makes clear. But there is no rationale for extending the relation to three terms, beyond the motive of obviating what seem legitimate, if troublesome, concerns. The tertiary reading is an ad hoc maneuver, in that it lacks any rationale independent of the disreputable motive of circumventing a criticism that deserves consideration.

Worrall's strategem is so commonly resorted to in philosophy that it is worth belaboring my opposition. Suppose that two people disagree about the artistic merit of a painting and cite different aspects of the painting in support of their positions. Each follows the other's reasoning and acknowledges that it does indeed lead to the position that it has been used to support. But neither allows that it *does* support that position, for each denies that the other's is the appropriate reasoning to use. It is of no help, in adjudicating this controversy, to relativize the judgment of each to the aspect that he emphasizes, maintaining that aesthetic merit is a binary relation between a work of art and a selected artistic feature. If this is what aesthetic merit is, then there *is no controversy to resolve*. There only *appears* to be a disagreement, because different features have been selected and the relational character of aesthetic merit has been concealed by the terms in which the differing views have been expressed. Even if both parties admire the painting, they are not truly in accord, because they are admiring different things about it. Indeed, it is not really *the painting* at all that they admire; this would require a further inference that could be blocked by other attitudes they have. Any claim to have solved the problem of how to rate the painting by this relational stratagem is sheer pretense; the problem has been *dissolved*. And yet it is clear in this example that the problem is real. The fact that the willingness of each party to credit the other's reasoning does not reconcile them shows that they are not operating with a relational concept of aesthetic merit.

The analogy can be improved. Suppose the painting's subject is a building, and that part of the painting's artistic merit lies in the architectural creativity that the building exhibits. The art critic wonders whether the artist invented the scene or was depicting a real building. Suppose there exists such a building, and consider two possible provenances for the painting, one in which the artist painted *that* building, and one in which, unaware of the building, he composed the scene independently. It is intuitive that in the latter case, the work is more praiseworthy. We may question whether, on this intuition, we have a *greater painting*, or only a more admirable *effort* by the artist. But our question is how to capture the intuition. We capture it by admitting the relevance of the artist's possible prior exposure to the building to our assessment of the work. We do not capture it by deciding that artistic merit is relational. On that approach, not only is no verdict as to the merit *of the painting* potentially at issue, but also it becomes impossible to ask whether how the painting was done should matter to its assessment; the very question has been ruled illegitimate. But this question does matter; critics dispute it and aesthetic theories give different answers.

The question of how well a theory is supported is not answered by a relative measure of the theory's support, given a particular provenance. On the tertiary reading, we never know that a theory is supported by the evidence, only that it is supported by the evidence *via* a certain provenance. What does knowing this enable us to conclude about the theory's credibility? Unless we weight provenances, examine all possible provenances, and calculate a net degree of support, the answer is *"nothing."* Yet, judging the epistemic warrant of theories is the very point of analyzing the evidential relation. Despite his defense of a use criterion over a temporal criterion, Worrall, like Gardner, is left with a skeptical position as to the epistemic legitimacy of weighting results that meet his criterion higher than others deducible from theory.

One philosopher whose theory of evidential independence purports to capture positive grounds for warranting theories is Ronald Giere.[33] Giere claims that to test a theory, a result must have a low probability of inducing us to accept the theory should the theory be false; and a high probability of inducing us to accept the theory should the theory be true. He argues that, in general, novel results satisfy these conditions better than other consequences of theories. In particular, if a result acted as a *prior constraint* on the construction of a theory, so that the theory was fashioned in a way that guarantees its conformity with the result, then the result cannot test the theory. A prior constraint has a low probability of leading to rejection of the theory *under all circumstances*; having functioned as a constraint, it has no potential to clash with the theory.

It follows that a result already deduced from an alternative theory will fail as a test. Its successful prediction from the theory to be tested need not, as I have put it, redound to this theory's credit, there being another theory equally entitled to such credit. Such a result cannot, therefore, carry a high probability of inducing acceptance of the theory to be tested, *whether this theory is true or not.* Thus, while Giere does not, in principle, impose a temporal condition on novelty and so does not require that novel results be surprising or unexpected, he is bound to require that they be *unexpected relative to alternative theories*, if their novelty is to qualify them to function as tests.

These observations do not, of course, constitute a positive analysis of novelty. We have at least some conditions that disqualify a result as novel, but we are not given sufficient conditions for novelty. Giere is prepared to argue cases, however, and we can extract a purportedly sufficient condition from them. His best case is the prediction of the diffraction pattern produced by a circular disk.

An impediment in the path of a beam of light produces a pattern of alternating light and dark bands of illumination, or fringes, at the boundary of the shadow cast by the impediment on a screen, and the pattern is characteristic of the shape of the impediment—a phenomenon known as "diffraction." This phenomenon was known by 1665,[34] and convinced Christian Huygens that light must have a

---

33. Giere, "Testing Theoretical Hypotheses," secs. 6 and 7.
34. Diffraction was evidently discovered by Francesco Grimaldi and reported posthumously in 1665.

wave structure that "bends around" impediments. According to Newton's *Optiks*, light has a corpuscular structure, requiring a different analysis of diffraction. Under Newton's influence, the corpuscular theory dominated the eighteenth century, but early in the nineteenth century, the wave theory was revived, first by Thomas Young. In 1816, Augustin Fresnel developed a wave mechanism for diffraction, based a upon a general, theoretical principle, which he called "Huygens principle": Points on a wave front beyond a diffracting object that disturbs an incident beam of light act as secondary sources of the disturbance, and the resultant wave at any point can be reconstituted out of the secondary waves. For this to be possible, the diffracting object must not affect the diffracted light, as it would in the corpuscular theory by exerting forces on the light corpuscles; its influence is only to block part of the incident light. In addition to this theoretical picture, Fresnel had available to him the actual diffraction patterns produced by impediments with straight edges. But contrary to Giere's account, Fresnel did not rely on these empirical results; his method is perfectly general and equally capable of supplying diffraction patterns for the straight and circular cases.[35] In fact, Fresnel's method significantly improves on the quantitative accuracy of corpuscular-based predictions for the straight cases.

In an attempt to discredit Fresnel's theory, Poisson deduced from it that the diffraction pattern produced by a circular disk would show a bright spot in the center of the disk's shadow, a phenomenon unknown and unexpected on the basis of the corpuscular theory.[36] However, François Arago's experiment to test Poisson's

---

35. Because of this independence, the patterns produced by straight-edged diffracting objects might be thought as strong a confirmation of Fresnel's theory as the more celebrated circular pattern, unless one imposes temporal restrictions on novelty. The fact that the circular case was *not* originally given special weight over the (known) straight cases in evaluating Fresnel's theory indicates that no temporal condition on the importance of experimental support was operative. On the other hand, there remains an important difference, having nothing to do with temporal restrictions, between the support provided by the straight and circular cases. The corpuscular theory was able to explain the diffraction pattern in the straight-edge cases, and I have argued that this fact ought to reduce the support they provide for a new, alternative theory.

John Worrall, in "Fresnel, Poisson and the White Spot: The Role of Successful Predictions in the Acceptance of Scientific Theories," in D. Gooding, T. Pinch, and S. Schaffer, eds., *The Uses of Experiment* (Cambridge: Cambridge University Press, 1989), uses the short shrift given circular diffraction to oppose temporal restrictions on novelty. But whatever the original reaction to the case of circular diffraction, it has come historically to bear a distinctively probative significance in support of the wave theory. Worrall attaches no importance to this development, dismissing it as the application of a mistaken, temporal view of novelty. In my (equally nontemporal) view, this development is fully warranted, and attributable to the fact that an alternative, corpuscular account was available for straight-edge diffraction, but not for circular diffraction. As to why the distinctive importance of the circular case was slow to be appreciated, many speculations are possible. But whether or not Giere is historically right to emphasize the confirmatory status of the circular case over the straight, he should be.

36. According to E. T. Whittaker, in A *History of the Theories of Aether and Electricity* (London: Thomas Nelson and Sons, 1951), p. 108, J. N. DeLisle had already discovered the bright spot in the eighteenth century, but that intelligence, if real, evidently failed to reach Fresnel, Poisson, or the figures of the French Academy, who awarded the 1918 prize for the diffraction of light to Fresnel's theory.

conclusion revealed the bright spot, unexpectedly confirming Fresnel's wave theory.

The condition that Giere's discussion of this case suggests is that if a result is unqualifiedly unknown, presumed nonexistent, and its contrary can be deduced from the best theory going, then it may be regarded as novel for a new, competing theory that predicts it. This condition is extremely restrictive, and it is to be hoped that something weaker will suffice, if only because of the counterintuitiveness of, and historical objections to, a temporal analysis. There is also a respect in which this condition needs strengthening: A result must differ qualitatively from known results that influenced the development of the theory for which the result is novel. I have noted that this requirement is not satisfied by stipulating that the result be unknown.

But there are also objections to which Giere's proposal is subject, to the extent that it does provide a positive analysis. One is that Giere's argument limits the dimensions of theory appraisal to acceptance and rejection. He does not consider neutrality or indifference, for example. He does not consider arguments we have encountered to the effect that the strongest epistemic stance warrantable, in principle, by any test is weaker than acceptance. A philosopher like van Fraassen, for example, would presumably respond that if a good test is what Giere proclaims it to be, then either there simply are no good tests, or no test can warrantably be judged good.[37] For there can be no test whose results make it probable that a theory is true, or entitle us to judge that its truth is probable.[38]

Another objection is that there is a difference between the conclusion that Giere actually defends and the conclusion that he needs. The conclusion he defends is that, at least in certain cases such as Fresnel's, we can be confident that a test will not lead us to reject a true theory, nor to accept a false one. For if Fresnel is right, Arago's experiment is likely to produce the bright spot, whereas if the corpuscular theory is right, this result is unlikely. The conclusion Giere needs is that if a test leads us to accept a theory, then the theory is likely to be true, and correspondingly, the theory is likely to be false if the test leads us to reject it. What we want, in other words, is that our tests be *reliable*, in that the decisions we base on them about theories are likely to be correct.

The difference might seem inconsequential. Assuming that if a theory is true, the test will likely so indicate, can we not infer that if the test does not so indicate then the theory is likely false? We cannot. What can be inferred is that if the test *is unlikely to indicate truth*, then the theory is not true. But the actual result of the test does not tell us what test results are likely or unlikely. So we cannot use

---

37. Van Fraassen uses the word 'accept' in such a way that acceptance does not carry conviction. One can accept a theory in his sense without believing it. He could take advantage of this idiosyncrasy to dispute the response I impute to him, but the point is unaffected. The utility of van Fraassen's sense does not survive his epistemology, and I shall not employ it. His epistemology is considered in chapter 6.

38. Giere claims that the relevant probabilities are physical probabilities, actual propensities in the phenomena. On this view, his critic should acknowledge that $pr(t/a)$ (in the notation introduced below) may well be high, but deny that we can have adequate grounds for asserting that it is high.

the actual test result to estimate the theory's truth on the basis of the conclusion Giere defends.

But can Giere not argue as follows? A result that leads us to accept a true theory is reliable; it does not mislead us as to the truth value of the theory. A result that leads us to reject a false theory is similarly reliable. So for a test that meets Giere's conditions, if the theory is true, the test is reliable, and if the theory is false, the test is again reliable. As we get reliability either way, we can deduce that the test is reliable without independent information as to the truth value of the theory.

The problem with this reasoning is that reliability has not been shown to be independently deducible both from the theory's truth and from its falsity. What is deducible from truth is not reliability, but the unlikelihood of a negative result. This consequence is consistent with unreliability, for it might be that a negative result is unlikely in any case. Similarly, what is deducible from the theory's falsity is not reliability, but the unlikelihood of a positive result, which could, so far as we are able to ascertain from the assumption of the theory's falsity, be unlikely in any case. If we cannot deduce reliability individually from truth or falsity, we cannot deduce it from their disjunction.

The difference between the conclusion reached and the conclusion needed amounts to a difference in the conditionalization of probabilities. Let 'a' represent a test result that leads to the acceptance of a theory, and 't' represent the truth of the theory. Then it is supposed to have been shown that $\mathrm{pr}(a/t)$ is high and $\mathrm{pr}(a/\neg t)$ is low. What we need to be able to conclude, however, is that $\mathrm{pr}(t/a)$ is high and $\mathrm{pr}(t/\neg a)$ is low. Bayes's theorem allows us to deduce what we need from what we have, but only in the extreme case where "high" means 1, or maximal probability, and "low" means 0 probability. In this case, the equation

$$\mathrm{pr}(t/a) = (\mathrm{pr}(t) \times \mathrm{pr}(a/t)) \div ((\mathrm{pr}(t) \times \mathrm{pr}(a/t)) + (\mathrm{pr}(\neg t) \times \mathrm{pr}(a/\neg t))),$$

a more general form of Bayes's theorem, reduces to

$$\mathrm{pr}(t/a) = \mathrm{pr}(t) \div (\mathrm{pr}(t) + 0) = 1$$

and we need not be concerned with the prior probability of the theory, $\mathrm{pr}(t)$, on the basis of antecedent knowledge. In other cases—the more plausible nonextreme situations—we would have to evaluate prior probabilities before concluding that a theory which a test result would have us accept is likely to be true. And Giere does not tell us how to do this, nor does he show that if we did do it the desired conclusion would then follow.

Mayo's analysis of the severity of tests suffers from a related shortcoming. Like Giere, she conceives of severity in terms of a test's reliability at discriminating true from false hypotheses. Specifically, she interprets high severity of a test of a hypothesis as low probability that the hypothesis will pass the test if it is false, where probability is the relative frequency of failures by the false hypothesis in a series of applications of the test. Her analysis requires a sequence of applications of the test within which we can independently determine the relative frequency of correct judgments of the falsity of the hypothesis based on it. Mayo applies this analysis to the data actually obtained in testing hypotheses. She formulates a criterion for the epistemic importance *of this data,* of the reliability of the inference from the data that test a hypothesis to the existence of phenomena that would support the hypothesis. Her analysis does not explicate this latter relation of support itself. For we judge the epistemic importance of empirical results for theories, if not the adequacy of experimental data to establish these results, without reference to the sort of testing sequences that determine relative frequencies. This judgment is plainly made in individual cases, where no such frequencies are available.

Indeed, often it is only because we first and independently judge that a conjectured empirical phenomenon would be epistemically important in theory evaluation, that we undertake to obtain data that will establish it. The question of confirmation theory, as practiced by the philosophers I have been discussing, the question that those philosophers have deployed varying conceptions of independence and novelty to address, is what such judgments are based on and whether they are sound. The answer cannot be that epistemic support depends on meeting Mayo's conditions for severity. No doubt, the answer does require a low probability that a result that supports a theory would obtain were the theory false. But that probability cannot be interpreted as a relative frequency of test results, even if Mayo is correct in so interpreting the probability that the experimental data used to establish the result would be obtained if an empirical hypothesis asserting the result were false.

Mayo need not impose a frequency interpretation in the general case of epistemic support. Her proposal can be that epistemic support requires severity in the sense of low probability, and that this probability is measured by a relative frequency when the hypothesis supported is an empirical hypothesis that experimental data test. The question then becomes what the basis is for the relevant probability judgments in the general case. This is the question novelty is intended to answer.

It does not seem that the gap in Giere's argument, or in Mayo's, is to be filled by historical information. The Fresnel case is a paradigm of novelty. Its claim to novelty does not turn on any of the points on which the different views of novelty that we have encountered conflict. Arago's experimental result was successfully predicted by Fresnel's theory, was unknown, was not involved in the construction of Fresnel's theory, was relevantly different from results that Fresnel did use, and was unexpected on the basis of—and even contrary to—alternative theories. If novel success does warrant imputing truth, or some measure thereof, to theories, or if, less ambitiously, it at least provides greater evidential support for theories than their other successful predictions provide, then Arago's result provides such warrant or support for Fresnel's theory. Why, then, do Giere and Mayo fail to establish this conclusion?

The problem, I believe, is their attempt to steer clear of the vexed question of the role of explanation in justifying a realist stance toward theory. They do not allow themselves the argument that the truth of Fresnel's theory is needed to explain how the theory was able to come up with the right pattern for circular diffraction. Nor are the other philosophers whose views of novelty and independence we have considered, and who are less inclined to realism than Giere, willing to invoke that argument. Without it, I suggest, realism will not be a viable position. That it can be made viable *with* an explanationist argument is yet to be shown. But before addressing that issue, we shall have to have in place an adequate analysis of novelty.

Constraints on the Analysis of Novelty

What conclusions should we draw from this discussion of extant conceptions of novelty? I have seven to propose, and shall express them in terms of a theory, T, and an empirical result, e, deduced from it. I call them "requirements" rather than "conditions," because they do not themselves constitute an analysis and are not to be confused with the conditions of an analysis.

1. RELATIONAL REQUIREMENT: Novelty is a binary relation between e and T.
2. ANTECEDENT KNOWLEDGE REQUIREMENT: e's novelty with respect to T is compatible with e's having been known prior to the development of T. The importance of explaining e may have been a reason for developing T, and it may have been foreseen that T would be unacceptable unless e could be deduced from it.
3. EXPLANATION REQUIREMENT: If e is novel for T, then T proposes an explanation of e, but no other available theory proposes an explanation of e.
4. USE REQUIREMENT: If e is novel for T, then e was used in at most an incidental or inessential way in developing T, so that T could have been developed essentially as it was even if e had not been available.
5. INDEPENDENCE REQUIREMENT: If e is novel for T, then e is significantly different from results on which the development of T depended, and also from results explained by alternative theories.
6. EPISTEMIC REQUIREMENT: If e is novel for T, then the explanation of T's ability to predict e successfully imputes some truth, at least approximate or partial, to T. To explain T's success, e must be considered confirmation of T, as providing some reason to think that T is at least approximately or partially true.
7. HISTORICAL REQUIREMENT: That e is novel for T is a historical, descriptive hypothesis subject to test, and the testing procedure cannot be expected to be more conclusive than tests are generally. There is no guarantee that novel status will be decidable in any given case. Nevertheless, there must be a clear indication of how novelty is to be decided in actual cases, and examples both of cases that clearly do involve novelty and of cases that clearly do not.

Although imprecise, these requirements are understandable in the context of the proposals and arguments that I have examined. An adequate analysis of novelty will at once implement them and increase their precision.

3

An Analysis of Novelty

Overview and Motivation

The basic idea of my analysis is to provide that a theory uniquely explain and predict an observational result without itself depending on that result for its content or development. The point of this independence is to ensure that there be no explanation of how the theory manages to yield the result other than to invoke the entities or processes that the theory posits. The explanation of the theory's explanatory success must be that the theoretical mechanisms it deploys are what actually produce the result. There must be no alternative on which the theory (purportedly) explains the result even if what really produces the result is unrelated to what the theory says produces it. Thus, the theory's proposed explanation cannot have been constructed in such a way as to guarantee the result as an outcome, whatever form that explanation took. If we are then to explain the theory's explanatory success at all, rather than to dismiss it as chance, we will have to attribute, in compliance with the epistemic requirement, some measure of truth to the theory.

The uniqueness of the theory's explanation prevents a competitor from preempting this claim to truth. If a competing theory also explains the result, especially if it does so with a similar independence of development, then chance becomes the better verdict. For unless the theories posit similar explanatory mechanisms, at least one of them must be successful by chance, and if chance operates anywhere, it would require additional argument to go beyond chance in interpreting any one theory's success. More generally, truth is not to be attributed to a theory in explanation of its explanatory success if the result explained can also be explained another way. Truth is the strongest hypothesis to make. If explanatory success supports this hypothesis at all, it does so only where the total explanatory situation leaves no viable recourse. The analysis of novelty thus has two parts: an independence condition and a uniqueness condition.

Uniqueness immediately introduces a temporal restriction. The possibility of competitors is always open, in principle, and the best case to hope for is that there be good reason not to expect them. One option is to index novelty to times, so that a result loses its novel status relative to a theory once a competing theory emerges to explain it. It would also be possible for a result not initially novel for a theory to gain this status through the independent disqualification of another theory's explanation of it. If new information renders the other theory's explanation unacceptable, the conditions for novelty with respect to the first theory would then be met.

I will avoid such complications within the analysis by making novelty depend on the absence of a viable competitor when a result is first explained and predicted, and by then dealing separately with the question of what effect the possibility or actual emergence of a competitor, or one's removal, has on the epistemic import of novelty. Novelty will then, in accordance with the relational requirement, be relativized to theories but not times, the temporal condition within the analysis obviating the need to provide for temporal changes in novel status. It should be noted, however, that the first explanation and prediction of a novel result need not coincide with the introduction of the theory to be credited with these resources. They may await the independent development of auxiliary information needed to deduce the result from the theory.

The independence condition poses more complex issues. I will have to advance some generalizations about the development of theories, and provide criteria for assessing the extent and nature of the involvement of empirical results in their development. The explanation and prediction of a result is normally understood to require a deductive relation between a theory and the result. This requirement must of course be qualified: Auxiliary information mediates the deduction; and what is strictly deducible is usually an empirical law from which the result is predicted, and which may itself be statistical. But the structure of the reasoning by which a theory is used to predict and explain is basically deductive. The relevant notion of independence must be compatible with the obtaining of a deductive relation, or the conditions for novelty become inconsistent.

Assumptions and Terminology

I shall be speaking in general terms of the novelty of a (observationally attestable) result O for a theory T. I shall assume that, as a matter of historical fact, T is a product of reasoning, and that this reasoning is, in principle at least, open to scrutiny. I do not deny the possibility that a theory could just occur to someone, in a dream or a flash of inscrutable inspiration. Indeed, the hypothetico-deductive model of testing and Reichenbach's distinction between the contexts of discovery and justification, because they, respectively, ignore and dismiss the provenance of theories, incline to this picture of the genesis of theories. But this is not, in fact, how theories originate, and the admiration we direct at the creative theorist would be misplaced if it were. If, in a given case, T does happen to be conjured up serendipitously, or if the details of its origination are unknowable, then my analysis will simply be inapplicable. That is, whether or not O is novel for T will then be indeterminate, as the historical requirement allows.

What I mean by insisting that reasoning produces T is that there are reasons for the assertions that constitute T, and that these reasons operate causally in the formulation of these assertions. The content of T is made to be what it is *because of* these reasons. Of course, many things may be psychologically influential. And part of the point of insisting on reasons is to locate T's lineage within the overall course of the development of science, emphasizing T's grounding in antecedent scientific knowledge over its subjection to cultural influences that act alike on science and other social institutions. But the main point is topic-neutral: Reasons are regarded critically; their causal role is mediated by critical examination of their importance and suitability. For a cause to be a reason, it must be regarded, by whom it influences, as an appropriate object of critical evaluation, possibly to be found wanting.

To regard T as the outcome of a self-conscious process of ratiocination really makes sense only on a propositional view of the nature of theories. Theories are systems of propositions that bear truth value. The propositions may be expressed mathematically, but the terms in which they are expressed are as semantically fixed as any terms in natural language. A theory is not expressible in uninterpreted, formal language to which divergent interpretations are variously assignable; at most its syntax is thus expressible.

Other views are possible, but to the extent that they are plausible, I regard them as variants of, rather than rivals to, the propositional view. It is fashionable, for example, to identify theories with sets of models. (This works especially well for geometric theories, and it is no accident that neopositivists, for whom formal systems have always been the paradigm for theories, incline to this view.) One then requires an additional hypothesis to connect the theory with a real physical system, and this hypothesis, rather than the theory itself, becomes the locus of propositional attitudes. It is a hypothesis to the effect that a given physical system is (or is isomorphic to) a member of the set of models with which the theory is identified that is doubted, believed, confirmed, or evaluated.

I prefer to let the theory itself be the object of propositional attitudes, and to inquire as to the reasons for proposing or believing the theory, rather than for supposing it to subsume some independently specified physical system. There is then no bracketed question as to the theory's own origination, no arbitrary linguistic restriction to observables in specifying physical systems, and no tendency to imagine, mistakenly, that whether or not a theory applies to a given physical system can be settled definitionally. *In practice*—that is, in application to examples—the semantic view of theories as defined by their models is parasitic on the propositional view, because the only way to specify nonarbitrarily *which* models define a theory is by reference to its propositional content. To say that a theory is *nothing other* than a set of models is a bit like saying that addition is nothing other than a set of ordered pairs of sequences of numbers. We cannot pretend that there is no further question as to why it is *these* pairs and not others, a question answerable by reference to the idea of adding. Or, if we can pretend this in the case of arithmetic, by treating it as a formal theory, we cannot in the case of physical theories.

The main motivation for a model-theoretic treatment of theories seems to be a concern that too close an association between a theory and some particular formulation of it in language will raise problems of individuation. Formulations in different languages are formulations of the *same* theory only relative to some scheme of translation. Philosophers impressed by the doctrine of underdetermination discern an irremediably arbitrary element in the adoption of any scheme of translation. W. V. Quine, for example, rejects the relations of synonymy on which translation depends. In effect, any identification of theories expressed in different languages is itself a theory for Quine; there is no guarantee of its truth, and its falsity is strictly compatible with any body of empirical evidence.

Add to this the doctrine, due primarily to Kuhn's *The Structure of Scientific Revolutions*, that different theories in effect constitute different languages, any comparison of which amounts to an act of translation, and we get a total breakdown of relations among theories. Kuhn argues that *theoretical terms*, terms for which there are no observational or operational criteria of application, are definable only implicitly, and possess determinate meanings only within the context of the assumptions and methods proper to a given theory. A term that appears in different theories cannot, then, be presumed to carry the same meaning in those different occurrences, and will be similarly understood by proponents of the different theories only to the extent—highly limited, according to Kuhn—that they are able to suspend their differing theoretical commitments and think atheoretically. If *all* thinking about the world is mediated by theory, the very language in which one thinks being irremediably infected by theoretical presuppositions, it becomes impossible for proponents of different theories to address common problems, to debate or even to understand their differences. Even terms that purport to describe universally accessible, observable conditions carry theoretical presuppositions—are "theory-laden." Theory choice cannot, then, depend on rational deliberation over the alternatives, on the weighing of common evidence, but must be attributed, by default, to psychological, rather than epistemological, considerations.

Since this challenge to the rationality of scientific decisions and to the objectivity of progress in science arises entirely from questions about how theoretical terms acquire meaning and how meanings, thus acquired, may be compared, one might think to meet it through an analysis of theories that is free of linguistic restrictions, as per the model-theoretic approach. My own response would be, instead, to dispute the doctrine of implicit definition and to deny that theoretical presuppositions ineliminably mediate all thought about the world. I do not share the motivation for the model-theoretic view.

I note that the worry occasioned by implicit definition concerns the mutual intelligibility of competing theoretical perspectives. Kuhn deduced that scientists educated in different traditions would lack a common language. The attempt to adjudicate differences objectively would fail *for that reason*—a result he called "incommensurability," meaning, in essence, that rival theories are subject to no common measure of success or standard or evaluation. The problem of individuation that exercises the model-theorist is, rather, the problem of recognizing the same theory under different linguistic formulations. The model-theorist hopes to identify theories independently of *any* linguistic formulation. But what were supposed to impede commensurability—the neutral comparison of rivals—were differing background assumptions that rivals could not help but make. Nowhere in arguing for the pervasiveness or entrenchment of such assumptions has a case been offered for supposing them indexed to languages. No reason has been given to think that theoretical convictions divide up according to the language the scientist uses. Differences in convictions, by creating different contexts for implicit definition, are supposed to *lead* scientists to speak differently; they are not themselves occasioned by differences in modes of speech. Theoretical presuppositions should therefore be identifiable independently of the semantic bifurcations that the incommensurabilist alleges them to cause.

If this is so, incommensurability gives us no call to prefer the model-theoretic view to the propositional view. We are free to adopt the propositional view, and thus accommodate the rational provenance of theories, without troubling over how serious—or real—a problem incommensurability is. Admittedly, incommensurability would, if real, challenge scientific rationality; but it is the rationality of comparative theory evaluation—the objectivity of grounds for theory preference—to which it pertains, not the rationality of theory provenance.

Now, reasoning is, in principle anyway, reconstructible. The reasoning that produces a theory can be represented as an inferential sequence of steps in an argument that connects the theory with the antecedent state of background knowledge from which it emerges. In practice, a reconstruction will be somewhat idealized. It will not contain everything causally influential in producing the theory, and some of its steps or inferential transitions may be more economical than those of the pattern of inference that was actually undergone. A reconstruction represents, but does not duplicate, the creative thought of the theorist. It seeks to capture what, in retrospect, prove to be the pivotal points at which a successful direction of development was initiated, whether or not their importance was appreciated at the time, while suppressing the peripheral or abortive. The same reasoning is variously reconstructible, with different points emphasized according to one's evaluation of how progress was achieved.

I shall call a reconstruction of the reasoning producing *T adequate*, if it presents a case sufficiently compelling to make *T* worthy of serious consideration, to motivate proposing it. An adequate reconstruction explains not only how *T* was reached, but also why one would regard it as a significant contribution. To some extent, the adequacy of a reconstruction is evaluative; the notions of "worth" and "significance" to which I appeal carry epistemic value. But at issue here is the provenance of *T*, not its evaluation once proposed. How *T* stands up to empirical test does not enter into the reconstruction. Rather, the adequacy of a reconstruction establishes the appropriateness or importance of *subjecting T* to test; it creates an onus or challenge to determine whether or not *T* is an acceptable theory. A reconstruction's adequacy is evaluative to the extent that the reasoning reconstructed is itself justificatory, independently of the subsequent stage of testing.

To make this idea more precise, I will suppose that among the propositions constituting *T*, some are *essential* to *T*, in that their rejection would constitute rejection of *T*. Any given presentation of *T*, especially one that seeks to apply *T* to some independent problem, will include many assumptions and conclusions that are subject to revision or replacement as part of a process of developing or refining *T*. Proponents of *T* could, consistently, disagree or change their minds about the legitimacy of these propositions; *T* is not committed to them. But, I am supposing, *T* is committed to *something*. A theory, on my propositional view, is not a set defined by its elements, but there are some propositions definitive of it. These I call the *basic hypotheses* of *T*. They are to be distinguished from those further hypotheses whose revision or replacement change *T* but do not reject or replace it. Such a contrast in status among the components of theories is necessary to provide for the individuation of theories while treating them developmentally. If a theory is identified definitionally with a set, only its application can be viewed developmentally.

The identification of basic hypotheses may itself have a retroactive, or reconstructive, element. What we take to be thus basic depends on how we interpret debates involved in the development and testing of theories. Claudius Ptolemy thought that astronomy required a physical mechanism for the transmission of circular motions among celestial spheres. The *Ptolemaic theory* of medieval astronomy, however, contented itself, for the most part, with geometric models of celestial motions without regard to physical realizability. Ptolemy himself would have laid no claim to what some latter-day astronomers called "Ptolemaic theory," not only because of the inclusion of artificial devices for fitting growing observational knowledge of planetary motions but also because of the omission or attenuation of key mechanistic assumptions. Maxwell originally thought that a mechanical analysis *of some sort* of the ether was necessary to his electromagnetic theory. Yet hypotheses as to the physical structure of the ether took on an optional character and were not to survive as ingredients of Maxwell's theory, which came to be identified simply with a certain system of equations.

One can write down Maxwell's theory very concisely, ever more concisely in subsequent formulations of the mathematics. Still, the theory is not *just* the formal equations; some interpretation is understood. The equations are no less propositional for their formalism. Similarly, Newton's mechanics is represented by his laws of motion and gravitation; it can be written as four equations, but equations that express propositions about motion and force. Newton himself, as he developed the theory, would have included much more. So would we: No theory attributing physical significance to spatial or temporal location as such is Newtonian, for example. But minimally, these equations count as basic hypotheses of the developed theory. Lorentz's electrodynamic theory is committed to the hypothesis of electroatomism, among others. That hypothesis, which, in effect, introduces what we have come to recognize, independently of the welter of theories employing it, as the electron, was crucial to distinguishing Lorentz's approach to electrodynamics from other coeval approaches.

I am supposing that given a developed theory, before the scientific community for evaluation, or historically circumscribed and subject to comparison with alternative theories, its basic hypotheses are identifiable independently of judgments of its successfulness. Those who assess the theory differently are in accord as to what propositions, minimally, they disagree about.

Now, more precisely, the adequacy of a reconstruction provides that there be a logically valid deduction of the basic hypotheses of T from premises that have some objective form of support. That is, one can produce arguments for the premises of the deduction, arguments that ultimately appeal to experience or to pragmatic advantages. I have three types of premises, distinguished by type of support, in mind: generalizations of suggestive experimental results; abstract appeals to simplicity or parsimony; and methodological prescriptions as to what type of hypothesis to invoke or avoid in certain kinds of situations.

An example of the first type is Galileo's law of falling bodies, or, more generally, the principle of the equivalence of inertial and gravitational mass. Another example is Einstein's generalization, from the failure of first-order experiments to detect absolute motion, that no empirical phenomena identify a state of absolute rest. These principles have the status of premises in reconstructions of the theories of relativity.

A pragmatic principle of simplicity might, for example, select a particular equation to codify data that, strictly speaking, admit of different functional representations, as when we connect data points (or their neighborhoods) by a smooth curve. A rather different example is Alexander Friedmann's assumption, basic to modern cosmology, that the universe looks essentially the same on a large scale from all positions in it, that this is not a special feature of our position. Attributing a unique centrality to our vantage point is empirically possible but needlessly complex and inelegant, in that it calls for some explanation as yet unknown.

Rejection of ad hoc hypotheses, and the introduction of new particles in connection with fields needed to effect symmetry constraints, are methodological prescriptions, though of very different kinds. The first discourages while the second encourages—indeed mandates—a certain hypothesis. The first is very general, appealing to the overall record of theories that rely on ad hoc modification to evade experimental problems. The second belongs to a particular successful program for understanding fundamental forces within quantum mechanics.

A more specific example of this second type is the assumption of a particular mathematical symmetry, the gauge group $SU(5)$, say, to unify the electroweak force with the strong force. Although the theorist has no empirical reason to expect this to be the right symmetry to assume, he has powerful reasons for beginning with *some such* symmetry and letting the outcome test his choice. He selects the mathematically simplest one consistent with his empirical knowledge, and would not be at all surprised if it should turn out to be the wrong choice.[1] What would surprise him— indeed, disrupt his whole program—would be the failure

---

1. In fact, the failure to detect proton decay is held to require a much more complicated symmetry group. See chapter 7.

of the strong and electroweak forces to merge under *any* such symmetry. For the empirically successful combination of the weak and electromagnetic forces is good reason to regard electroweak theory as a special case of a more inclusive symmetry. It is good reason, that is, to propose such a symmetry even if it is not good reason to expect any particular choice to be confirmed. The cogency of the reasoning that produces T consists in establishing it as worthy of consideration and test; it is not supposed to warrant acceptance, as the premises, though nonarbitrary, are defeasible. Such cogency is what the adequacy of a reconstruction amounts to.

The premises of the deduction of T's basic hypotheses may be tentative or controversial, as any topic-specific methodological sanction is likely to be at some stage. Experimental generalizations may even admit of known exceptions. Yet the theorist has some reason, whether or not others share it, to place at least heuristic confidence in them. Adequacy cannot be soundness; it is just rationale.

The stipulation that a reconstructive argument be deductive is a convenience, there being nothing that one can assume about inductive arguments as such. The inferences reconstructed need not be formally deductive. Once one understands the reasoning, one can formulate it deductively by a judicious choice of premises, ampliative inferences being recast as generalized assumptions. There is no question of imposing a tight deductive structure on historical processes of theorizing.

Still, the stipulation that reconstructions be deductive may alarm an intuition that expects novelty, on any account, to turn out ampliative. It is helpful here to note that once one gives up temporal constraints on novelty, as my antecedent knowledge requirement has done, the concept of novelty can no longer be expected to respect the term's psychological connotations of surprise, abnormality, or originality. The term carries these connotations in its scientific use because *one* abiding interest in novelty is an interest in the *growth* of empirical knowledge. One wants to know when an empirical result *adds* to beliefs about the world. As explained in chapter 1, mine is a different interest pertaining to theoretical belief, not to its growth but to its sanction. I am concerned with the ability of an empirical result to provide epistemic warrant for theory. Accordingly, it is an epistemic, rather than a psychological, sense of 'novel' that I seek. This epistemic sense captures *part* of the concept of novelty operative in scientific practice, omitting connotations that speak to ancillary interests. This is the way of technical definition in general.

A reconstruction is *minimally* adequate if it is adequate and, first, no empirical premise of it—no premise of my first type—may be removed or replaced by a logically weaker proposition without ruining its validity; and, second, the conjunction of its premises is not logically simplifiable. In other words, a minimally adequate reconstruction assumes nothing more about empirical results than is strictly necessary to generate the theory via the basic line of reasoning used, and it includes no logical redundancies, nor statements whose deletion would leave its deductive power unaffected. Accordingly, if S is an empirical statement that implies but is not equivalent to S', and a reconstruction is minimally adequate with S' as a premise, then it is not minimally adequate with S as a premise. If S and S' are each premises, then minimal adequacy is not preserved by replacing S by

'S or ¬S''. The simplicity of the conjunction of premises of course requires that each premise individually not be simplifiable. So if S is a premise, then, for example, 'S and W', where W is a necessarily true statement, and 'S or U', where U is a necessarily false statement, are not.

The idea is to circumscribe novelty by identifying what is essential to the reasoning used to generate the theory, specifically, with respect to the role of empirical results. But, again, 'essential' is to be understood in a logical, rather than a psychological, sense. It might be that the theorist would not, as a matter of psychological fact, have thought of the theory unless he knew of a certain empirical result, although the theory could have been validly deduced from other knowledge he had and assumptions he made. Then the result will not figure in a minimally adequate reconstruction. It might be that the result had a crucial role but was itself incidental to that role, a different result, had it been available, serving as well. Then a statement of the result actually used is replaceable by something logically weaker without detriment to the cogency of the reasoning; namely, by the disjunction of statements of results that would, individually, have served. The disjunction alone will figure in a minimally adequate reconstruction, although only one of its disjuncts was actually entertained by the theorist. A result that has no role in the actual reasoning may thereby be represented in a minimally adequate account of what is "essential" to that reasoning, in the nonpsychological sense. In short, a result that motivated the theorist may be inessential to his reasoning, while one never contemplated is essential.

I turn now from O's role in the reasoning that produces T to its observational character. O is not to be understood as a particular act of observing, nor as a report or description of such an act. Rather, O is the phenomenon observed, and, as such, may be reidentified across any number of observations. Different observations, under different conditions and using different equipment, may all count as observations of O. Nor need there be any commonality to the physical objects under observation, nor to the mechanisms for producing O, in all cases where O is observed. Thus, O involves a significant element of abstraction; it is a type having instances, and the identification of something as an instance of O is, in principle, defeasible. It is not guaranteed simply by the character of observational experience. In a given case, it might be a matter of debate whether or not what is being observed is really (an instance of) O. In general, what makes something O, what must be determined or corroborated to settle such debates, is satisfaction of a description that involves theoretical commitments, rather than just a description of visual qualities. An account of what O is, whereby it can be identified across different cases of observing it, will typically cite some process that can occur in different ways and can involve different physical entities. O is observed only when a general understanding—in principle, defeasible—of this process has been reached.

In scientific parlance, O is an "effect." Its significance depends on identifying that of which it is an effect, the causal process of which it is the outcome. An example is stellar parallax, an effect of the earth's motion. Nearer stars are displaced against the stellar background as the observer is carried around the sun. Whether there *is* such a phenomenon, and whether any particular sighting of stars was a case of observing it, were speculative and controversial questions over a substantial historical period. Once Nicolas Copernicus set the earth in motion, the answers became crucial to the theory of the structure of the universe. To reconcile the failure to observe parallax with the earth's orbital motion required so vast an increase in the dimensions of the heavens as to render hopeless any physical model for the communication of celestial motions in a universe structured on rotating spheres. In the absence of an alternative theory of celestial motion, the failure to observe stellar parallax discredited the hypothesis of a moving earth. Parallax was first observed telescopically in the late nineteenth century, after Newtonian theory had eliminated its importance as a test of this hypothesis by showing that it would not be observable under earlier conditions. Parallax does not figure in a minimally adequate reconstruction of Copernicus's reasoning, and its observation would have proved an important confirmation of Copernicanism had the hypothesis of a moving earth still been in question.

Another example is the deflection of starlight passing the limb of the Sun, an effect of the Sun's gravity. This effect is also absent from a minimally adequate reconstruction of the reasoning behind the theory that identified its cause. And its observation by the Eddington eclipse expedition of 1919 was a major confirmation of general relativity, which was very much in question. Similarly, the bright spot in the center of the circular shadow cast by a diffracting object, which was observed in Arago's experiment, is an effect of the wave nature of light, and proved an important confirmation of the wave theory. This effect is likewise absent form a minimally adequate reconstruction of the provenance of Fresnel's theory. The increase of mass with velocity, first observed in accelerator experiments in the 1930s, is an effect of the inertia of energy. Its observation would have been significant on behalf of special relativity had not the generalization of this theory already been well confirmed. This effect, too, is absent from the provenance of special relativity, properly reconstructed.[2]

O involves generalization, but is now itself to be generalized. In an actual case, a description of O will relate it quantitatively to its cause. The amount of mass increase is a function of velocity. The amount of parallax depends on the proximity of the displaced star and the diameter of the earth's orbit. A theory that explains the process behind an effect will give a quantitative prediction of the effect, and confirmation will depend, in general, not just on the realization of the effect itself, but also on the quantitative accuracy of the prediction. General relativity, for example, predicted not just that starlight would be deflected, but also that it would be deflected by roughly twice the amount to be expected on Newtonian gravitational theory. The next generalization suppresses this dimension of O.

By the *qualitative generalization of O*, I mean *the effect itself*, the type of phenomenon—itself a type—that O instantiates, independently of considerations of quantitative accuracy. This need not be unique. Under some minimal description, an effect might subsume distinguishable subeffects. Stellar parallax due to

---

2. See chapter 4.

the earth's motion may be regarded, in certain contexts, as but one type of a more general parallactic effect. We must provide for the possibility that O is multiply, qualitatively generalizable, depending on how the effect is specified in a given context.

One gets O's qualitative generalization by eliminating from O's description all numerical values predicted from T and compared with measurement in using O to test T. To be rigorous, numbers should be replaced by variables which then require existential generalization and restriction to qualitatively fixed ranges; for example, "measurable," "observable," or "significant" ranges. Instead of saying that starlight will be deflected by such and such an amount, we will then say that it will be deflected by a measurable or observable amount. In practice, qualitative generalization can be simple deletion, from, say, "the pointer will be deflected by ten degrees" to "the pointer will be deflected."

The reason for qualitatively generalizing O is to provide a criterion for counting different predictions, especially predictions from different theories, as predictions of the same effect. If the predictions disagree only with respect to quantity, then it is the amount of the effect, not the effect itself, that distinguishes them. The same effect is predicted if predictions of O are alike under qualitative generalization. This criterion equips the analysis of novelty with an unusually exacting standard of predictive uniqueness.

The motivation for such a standard is the leeway available to theories in accommodating quantitative discrepancies with predicted values of an effect. There is simply too much opportunity—in the form of optional developments beyond a theory's basic hypotheses and appeal to auxiliaries—to bring theories into conformity with the data, for an epistemically probative role to be assumed for quantitative accuracy alone. Any number of examples may be produced in which flexibility in the addition of hypotheses and the use of collateral information turned an initial anomaly into an apparent confirmation. Nineteenth-century astronomers like George Airy revised estimates of the masses of known planets to correct the orbit of Uranus as predicted from Newtonian gravitational theory. They hypothesized additional matter inside Mercury's orbit to correct Mercury's precession. General relativity's predictions for the gravitational deflection of starlight changed with different versions of the theory. Various hypotheses compatible with Newtonian theory were proposed to augment the amount of deflection beyond what that theory predicted, so that no difference would remain with predictions from the final form of relativity. Even now there is a question as to the ability of Arthur Eddington's original results to discriminate relativity's predictions from Newtonian ones. There are also challenges to the appropriateness of the Schwartzchild approximation used in applying relativity in this case, with the possibility of obtaining different predictions from the theory with different solutions to the field equations.

Such resources of application are typical of the more important and successful theories in physics. Whether they are to be decried as epistemic weakness or commended as methodological strength depends on what is at stake. As often as their exploitation has prolonged a theory's longevity beyond its merits, it has

produced major discoveries that an unforgiving adherence to popular methodological norms would have missed. The reluctance to reject theories is as important to the achievements of science as the falsifiability of theories is to the
objectivity of science. But this acknowledgment does not counsel reliance on
quantitative agreement when the issue is the epistemically probative status of a
result for one theory in a case where a rival theory explains and predicts the same
effect.

The suppression of quantitative considerations via the qualitative generalization of O is intended to make novelty hard to achieve, so that reluctance to
accord it epistemic weight will be correspondingly weakened. A more liberal conception that allows quantitative accuracy to meet the uniqueness condition for
novelty is possible and defensible for some purposes. But my agenda of warranting
the attribution of truth to theories requires a more stringent interpretation. It will
be important to keep this motivation in mind. Quantitative accuracy is of great
importance in fundamental science, and theories may be judged unacceptable
due to very small discrepancies between the amount of an effect that they predict
and the amount measured by experiment. Holding theories to an exacting standard of accuracy reflects the stringency of conditions for belief or acceptability.
But the stringency I seek requires the *elimination* of quantitative considerations
from the analysis of novelty. Novel status cannot depend on the quantitative comparison of the predictions of rival theories, because this comparison inherits the
instability of auxiliary information used to make predictions. A result novel for a
theory is to be one that no rival theory predicts *at all*, even inaccurately. Thus,
novelty can fail because of the qualitatively correct but quantitatively incorrect
predictions of a rival theory.

This move does not weaken the stringency of conditions for belief. Instead, it
means that the prediction of novel results is insufficient for belief, which will
further require that the novel results be predicted with quantitative accuracy. Accuracy is part of what it takes for the prediction of novel results to be *successful*
prediction; inaccurate predictions are unsuccessful. The point is that this is an
additional requirement, not to be included among the conditions for novel status.
Accuracy is required for the success of predictions, not for their novelty.

An example to illustrate the difference is Einstein's analysis of Brownian motion. In 1905, Einstein deduced from kinetic-molecular theory that microscopi-
cally observable bodies, suspended in a medium, are subject to measurable motions not attributable to motions of the medium. The phenomenon had been
observed by Robert Brown as early as 1828, first with pollen particles suspended
in water. It was not immediately clear whether the movements Einstein predicted
were identifiable with Brownian motion, but whether known in advance or not,
they constituted an effect for which atomic theory provided the first and only
explanation. This is crucial for the novel status of the effect with respect to atomic
theory. But what *confirmed* atomic theory, what convinced such skeptics as Wilhelm Ostwald of the reality of atoms, were detailed quantitative results obtained
by Jean Baptiste Perrin and Theodor Svedberg during the period 1912–14. Perrin
produced, by centrifuge, uniform latex corpuscles of known dimensions whose

measurable diffusion rates matched those deduced by Einstein. Einstein called Perrin's results "the first complete and absolute proof of the formula."[3] The interval between the explanation and prediction (1905) and the demonstration of its quantitative accuracy (1914) highlights the difference between the novelty of the effect predicted and its confirmatory status.

The distinction between quantitative and qualitative interpretations of the uniqueness condition is admittedly slippery in the abstract. An apparent difference in kind between the predictions of two theories, indicating that the predicted results instantiate distinct effects, may be reconstrued as a difference in degree to make it appear that the same effect is predicted. For example, the nonexistence of an effect can be represented mathematically by assigning a zero value to its measure. However valuable and appropriate this may be as a tactic in economizing a theory's presentation and its application to technological problems, it is a logical legerdemain that should not be allowed to obviate the conceptual difference between predicting an effect inaccurately and not predicting it at all. The difference is real between a debate over the existence of a natural phenomenon and a debate over its quantity.

The Lorentz electron theory was committed to an influence of the earth's motion through the ether on the velocity of light; it did not avoid this commitment by arranging, through the use of G. F. Fitzgerald's hypothesis that such motion contracts material bodies, for the measure of variation in the velocity of light, with different orientations of bodies with respect to the earth's motion, to be zero. It was clear that the effect was still supposed to be real, if undetectable. And it became clear, although this took a while, that special relativity was fundamentally different in denying the effect altogether, even if the difference was empirically unattestable. The significance of the null result of the Michelson-Morley experiment was not simply a matter of degree, and Michelson could not make it so by lamely reporting the result as "less than the twentieth part" of the expected value.[4]

It is similarly implausible to represent the significance of a zero cosmological constant as a matter of degree. The fact that the cosmological constant is precisely zero—so far as we have determined (which is very far, indeed)—is understood to require a special explanation. It has become a test of certain theoretical proposals in contemporary cosmology—involving constraints on ways of combining topologically disjoint regions of space-time—that they provide such an explanation, or at least leave in tact such explanation as we already have. Were the cosmological constant found to be positive, it does not seem that there would be any corresponding requirement that its precise value be derivable from fundamental theory. There would only be a need to explain why its value should be positive at all—

3. Albert Einstein, *Investigations on the Theory of the Brownian Movement* (New York: Dover, 1956), p. 103.

4. A. A. Michelson and E. W. Morley, "On the Relative Motion of the Earth and the Luminiferous Aether," *Philosophical Magazine*, 5 (1887):449.

that is, why there should be a gravitational effect unassociated with the distribution of mass-energy.[5]

A more familiar example is the textbook treatment of Newton's laws of motion. Mechanics achieves a certain elegance and economy by representing the principle of inertia, Newton's first law, as a special case of the second law with zero acceleration. But conceptually, the first law is not just an instance of the second. The second tells how objects respond to an impressed force. The first describes the force-free situation and is definitive of the concept of a "natural state," a condition that Newton had to reconceptualize. Newton was right to distinguish them, even if a condensed format is suitable now.

Given the notion of O's qualitative generalization, we are prepared to classify certain cases, which have been historically influential in theory evaluation, as cases in which rival theories predicted the same effect. And we can formulate a sense of 'novelty' that is epistemically more robust than fidelity to historical practice alone demands, robust enough to withstand challenges from normative philosophy.

Conditions for Novelty

My two conditions for the prediction of an observational result O to be novel for a theory T may now be stated:

Independence Condition: There is a minimally adequate reconstruction of the reasoning leading to T that does not cite any qualitative generalization of O.

Uniqueness Condition: There is some qualitative generalization of O that T explains and predicts, and of which, at the time that T first does so, no alternative theory provides a viable reason to expect instances.

The conditions require that no qualitative generalization of O be involved in the reasoning that produces T, and that at least one such generalization be supported by T but by no viable alternative theory to T. These conditions may be unpacked into more self-contained, if less manageable, form, as follows:

1. There is a valid deduction D of the basic identifying hypotheses of T from independently warranted background assumptions.
2. With respect to the specification of empirical results, the premises of D may not be logically weakened without ruining D's validity.
3. The conjunction of the premises of D is not logically simplifiable.
4. D contains no description of O; nor does it cite any more general description A that, by deleting from O's description numerical values pre-

5. The theoretical situation in this example is far from clear. At the same time that there are reasons in evolutionary cosmology to expect a cosmological constant near zero, there are reasons in elementary particle physics to expect a high energy density for the vacuum, to which the cosmological constant is proportional. This complexity does not affect my use of the example.

dicted from T and compared with measurement in using O to test T, abstracts from O the type of effect or phenomenon that O instantiates.
5. There is some A, A_0, of which T, by explaining and predicting O, provides good reason to expect instances.
6. No alternative theory to T provides a viable reason to expect instances of A_0.

It might help legitimate the cumbersome character of the qualifications in this analysis to emphasize, now more directly, that the conditions are intended to be collectively *sufficient* for novelty. They are not intended as necessary conditions of a conceptual analysis, for that would serve no epistemic purpose. Novelty outside these conditions is simply not at issue. I shall, in the ensuing discussion of examples, deny the novelty of certain results, but my meaning will be that they fail to qualify as novel on the basis of my conditions. The case I wish to make for the justification of theories on the basis of their successful novel predictions will assume that these conditions are met. I claim no justification for theories under other conditions that might, independently, be thought to qualify a result as novel, or that reflect wider or looser uses of the term than mine. To impart appreciation of qualifications that remain even in the expanded form of the analysis, I will compare two of its applications: the light-deflection and Mercury perihelion tests of general relativity.

In both cases, Newtonian gravitational theory provides an alternative explanation to Einstein's of the predicted effect, in apparent violation of the explanation requirement for novelty. In each case the differences between the predictions may be represented in purely quantitative terms. That is, each theory, augmented by appropriate auxiliaries, predicts the effect; neither takes it to be nonexistent or zero. And the effect predicted is the same, qualitatively generalized. Assuming the relevant auxiliary information that was generally accepted at the time the effects were used to corroborate relativity, relativity's predictions are quantitatively successful and those of Newtonian theory are not. The Newtonian estimate of light deflection is off by 100 percent, and there is a persistent excess in the measured precession of Mercury that Newtonian theory needs independently unmotivated and largely discredited auxiliary hypotheses to cover. For example, intermercurial matter was not discovered, and to suppose it to remain hidden by Mercury or the Sun violates Newtonian dynamics. This difference does not affect the application of the analysis.

What does matter, what makes the verdict as to novelty turn out differently in the two cases, is that the use of Newtonian gravity to explain light deflection, at the time that relativity predicted it, is not viable. The problem is not that there is no viable way to make the prediction quantitatively accurate; there is no viable Newtonian explanation of the effect *at all*. This is because the independence of the velocity of light, a hypothesis well established by special relativity and firmly embedded in the prevailing relativistic mechanics of the time, is incompatible with a Newtonian treatment of light as consisting of material corpuscles subject to Newton's laws of motion under a gravitational force. Light corpuscles cannot change velocity in response to gravity conceived as a force of attraction; light can only change in frequency in response to a change of position in a gravitational field. In order to obtain, even inaccurately, a gravitational deflection of light from Newtonian theory, one must dismiss major theoretical developments since the time of Newton. The entire development of field theory in the nineteenth century, incorporating the wave theory of light into electromagnetism, must be discounted and a Newtonian corpuscular model resurrected. Then the role of light in relativistic mechanics must be discounted. And if one is prepared to do this — if one is willing to pick and choose one's auxiliary assumptions so as to produce the desired result without regard to what assumptions are justifiable with respect to current knowledge — then any theory can be credited with predicting anything it needs to predict, and the whole idea of testing becomes vacuous.[6]

Newtonian theory equates gravitational and inertial mass. If, on the basis of the equality of their measures, one identifies these forms of mass conceptually, one obtains the principle of equivalence; and on this basis alone, without introducing the effects of forces on light corpuscles, one obtains the deflection effect. For the equivalence principle implies that in a gravitational field light will behave exactly as in a uniformly accelerated reference frame. Since the path of a light beam must appear to bend from the perspective of such a frame, it must bend in a gravitational field. This is Einstein's famous thought experiment with the elevator. One might be inclined to credit Newton with an explanation of deflection on this basis. But the thought experiment is Einstein's, not Newton's. Newton does not identify the forms of mass conceptually; he does not propose a geometric analysis of force to underpin the equivalence principle conceptually. Accordingly, he cannot offer the equivalence principle as an explanation of deflection, even if his theory, as used (interchanging inertial and gravitational mass in his equations), can be made to yield deflection. Only insofar as he posits a corpuscular structure for light, and subjects the corpuscles to the ordinary dynamics of motion, does he have an explanation of deflection. And this special relativity does not allow him to do.

Although it is natural and satisfactory in many contexts to suppose that the ability to predict reflects the explanatory resources of a theory, the expression "explain and predict" is not redundant. And I have been careful, in formulating my analysis, to require not merely that T, for which O is novel, predict O, and that O not be predicted by an alternative to T; but also that T alone provide a viable reason for expecting the effect which O instantiates. Thus, light deflection satisfies the analysis despite the fact that Newtonian theory could be used to predict it.

By contrast, Newtonian theory is very much to be credited with an explanation and prediction of planetary precession. Precession as such, discounting its quantitative measure, is simply a consequence of the Newtonian insight that gravitation is universal. Only if a planet were attracted by the Sun alone would its orbit be closed. The upshot is that the deflection of starlight past the limb of the Sun qualifies as a novel prediction of general relativity, while the prediction of the

---

6. A general defense of epistemic constraints on auxiliaries is offered in chapter 6, under "Empirical Equivalence and Underdetermination."

precession of Mercury's perihelion does not—notwithstanding the fact that both results were historically influential on behalf of general relativity.

And, it bears repeating, the difference has nothing to do with the fact that the precession case antedated relativity while the deflection case was a new prediction. Had light deflection actually been predicted and given a Newtonian explanation when one would have been viable, then that effect would qualify as novel for Newtonian theory; it would also be novel for relativity, since when relativity explained it, the Newtonian explanation was no longer viable. Via this scenario, light deflection would qualify as novel for both theories, although we must disallow that it be epistemically probative for both at any one time.[7] But its simply having been observed in advance, without being explained, would not matter.

One might wonder about this. Suppose that a result is well documented, recurring reliably under well-understood conditions, but is unexplained. It is *predicted*, in that a strong expectation naturally arises as to its recurrence, but no *theory* predicts it on the basis of its own explanatory mechanisms. Later, a theory is developed that predicts and explains the result in a way that satisfies the independence condition. The uniqueness condition also looks to be satisfied, since the alternative, preexisting basis for predicting or expecting instances of the effect in question is not provided by a theory. Are we to count the result as novel for the new theory, or are we to fault the analysis for arbitrarily restricting the form of competition that uniqueness disallows to *theoretical* competition? After all, no analysis of theoreticity has been provided to exempt cases in which competition, coming from something less than a theory, is innocuous to novelty. Why does it not count as a "theory," so far as the analysis is concerned, simply to generalize the result, to affirm its recurrence universally under similar conditions? Such a commitment, outstripping any finite body of observational data, would count as "theoretical" on at least some ways of distinguishing observation from theory (Quine's way, for example).

The intention of the analysis is to invoke an established concept in speaking of "theory," to defer, in effect, to normal scientific usage. Philosophical usage, concerned to mark off directly observed facts, free enough of defeasible interpretation to serve as a foundation for inferential knowledge, is lamentably broader. Basically, as urged in chapter 1, scientific usage requires a theory to posit a physical structure in terms of which it explains and predicts some independently circumscribable range of phenomena—not necessarily uninterpreted—which there is independent reason to regard as subject to unified treatment. The proper place to defend the distinction that the term 'theory' embeds into the uniqueness condition is a discussion of the problem of empirical equivalence.[8] For there, it be-

---

7. That is, we must disallow this if, by "epistemically probative," we mean that a result contributes to warrant for theoretical belief. Some philosophers—for example, Larry Laudan—think that a result can be epistemically probative in a sense weaker than this but stronger than offering pragmatic support. I do not discern such an intermediate sense, and shall explain why in chapter 6. The general question of the epistemic significance of novelty is the subject of chapter 5.

8. See chapter 6, under "Empirical Equivalence and Underdetermination."

comes important whether or not the equivalent alternatives allegedly available to a favored theory are themselves theories.

Supposing, then, that the mere universalization of a result is not a theory, so that the uniqueness condition is satisfied and the result in the case at hand qualifies as novel for the new theory predicting it, one is entitled to ask why novelty should be so sensitive to this difference. The reason is found by recalling what the analysis is *for*. A novel result is supposed to contribute to the warrant for theoretical belief. It does so only if the beliefs to be warranted provide the sole reason to predict the result. Where there are competing grounds or reasons to expect the result, neither competitor can be warranted. For the warrant of each would defeat that of the other. Where theory T_1 is incompatible with theory T_2, a reason to believe that T_1 is true is a reason to believe that T_2 is false. And to the extent that there is reason to believe T_2 false, such support as there is for T_2 cannot warrant belief. If it is not a theory that produces the expectation of recurrences of a result, but mere familiarity—mere exposure to observational regularity—then although the result is expected, there is no *reason* to predict it. So there is no competing reason to defeat the warrant that a theory provides.

This is a Humean point. David Hume parlayed a rejection of the possibility of warrant for causal hypotheses into a general attack on inductive reasoning. The implications of Hume's results for realism will be discussed in chapter 5. In defense of the uniqueness condition, I am anticipating that discussion here by claiming that the observation of a regularity in itself is no reason to project that regularity onto new cases. Until there is some potential explanation of why the regularity holds, an explanation that identifies conditions on which it depends, there is simply no basis for deciding whether or not it should be expected to recur in a given case. It is expected; but, as Hume emphasized, this is a point of psychology, not reason. The question of warrant for the explanation that rationalizes the expectation is another matter. The claim is not that there must be a warranted explanation before there is reason to project a regularity, simply that there must be a (potential or purported) explanation (that is not independently discredited). What difference does it make that one has an explanation if there is not yet reason to think that the explanation one has is true? The answer I propose is that having an explanation is a reason to think that the regularity *is explainable*. And that is a reason to treat it as a real, if not as yet understood, effect, rather than as happenstance—which is to say, to project it. Of course, two explanations, both unwarranted, that counsel incompatible projections are no better than none when it comes to *what* to project.

There are two kinds of argument for disqualifying regularities, as such, as reasons for prediction. One cites paradoxes that arise for enumerative induction that require explanatory hypotheses to resolve. The fact that many A's have, without exception, been B's makes it no more likely that the next A will be a B than that it will be a non-B, absent some hypothesis as to what has made those A's B's. It is all too clear that my chances of surviving another day do not increase with the number of days that I have already survived; the longer I have lived, the less likely I am to continue living. The ramifications of inductive paradoxes will be important in chapter 5. The other argument, which I wish to make here, appeals to the fact that one would be loath to infer a hypothesis that offered no explanation at all of the facts from which one infers. Hypotheses are not inferred by enumerative induction alone, for the very good reason that enumerative induction by itself lacks the resources to discriminate among incompatible, alternative hypotheses. Further assumptions are always crucial to the inference, and what this requirement amounts to is that there be some explanation of why the events or traits to be projected should accompany one another. It is only by appeal to explanatory hypotheses that regularities can be nonarbitrarily projected, just as it is only by appeal to regularities that hypotheses can be evaluated.

Hume thought that this produces a vicious circle; we need not follow him that far. But we should reject as baseless an asymmetry that favors enumerative induction as warrant for hypotheses over explanatory hypotheses as warrant for the projection of regularities. The problem with unexplained regularities is that there are too many ways to project them, as well as the supposition—equally plausible *ab initio*—that they are not projectable at all. The problem with unwarranted explanatory hypotheses is that there are alternative potential explanations. In this situation we may acquiesce in skepticism as to warrant altogether, or treat both enumerative induction and explanatory power on a par in reckoning the warrant for hypotheses. To induce enumeratively while declining to abduce—that is, to allow explanatory power a role in supporting inference—is an arbitrary preference, and it will not survive the argument of chapter 5.

It is also a preference that discounts or ignores the way people actually reason. Induction and abduction are equally entrenched in practical reasoning, and necessarily so. This fact should give pause to an epistemological naturalist who concedes the need to justify one's justificatory standards. For where is such second-order justification to be found but in paradigm situations of established justificatory practice?

In issuing this admonition, I emphasize that it is *justificatory* practice that I mean to address. Not all cases of projection are justificatory. Incontestably, people do form expectations, and in that sense, "project," on the basis of unexplained regularities. It is the *defense of beliefs* about unexamined cases that requires explanation. There are examples of projection of unexplained regularities within scientific practice, and Newton even made it a "rule of reasoning" to presume that observed regularities hold in general pending evidence to the contrary. I think this should be taken as a *methodological* prescription as to how to proceed when however we proceed takes us beyond the evidence available, not as an epistemic claim. The justification of belief depends on at least adumbrating, however tentatively, some explanatory connection between the items whose co-occurrence is generalized. This condition will be made more specific in chapter 5.

My conclusion for now, anticipated by the antededent knowledge requirement, is that the uniqueness condition has nothing to fear from the fact that results it qualifies for novelty may be well known and readily predictable in advance of the theory for which they are to be novel.

4

Applications

I will revisit two examples in further detail, and initiate consideration of two others, to display the analysis of novelty in action. The purpose is to illustrate how decisions about novelty, based on the analysis, are to be reached in practice.

Fresnel's Theory of Diffraction

Though anticipated by Robert Hooke, the wave theory of light may be said to have originated in Christian Huygens's hypothesis of an ether through which light is propagated as a wave disturbance of constant, finite velocity. In the case of the transmission of light through a homogeneous material medium, such as water or glass, the ether is imagined to permeate the medium, and the velocity of light through it is related inversely to its refractive index. Young contributed the hypothesis that light travels in series or trains of waves that can interfere, producing the alternately light and dark bands that had been studied by Newton in connection with the colors produced by thin plates—a phenomenon known as "Newton's rings." Young was guided by an analogy of light to sound, and assumed a longitudinal wave vibration to account for the raylike behavior of light. He explained the bands or fringes produced in diffraction by supposing that light reflected from the diffracting body interferes with incident light. This theory was unable to provide detailed predictions for many known diffraction phenomena. By 1815, diffraction had been extensively investigated experimentally, and a wealth of unexplained information was available.

Fresnel, beginning in 1814, contributed a number of physical assumptions, some of which were adumbrated in chapter 2. In accordance with the "Huygens principle," Fresnel treated each point on a wave front as a source of secondary waves, the contributions of all of which he combined to determine the intensity of the light. He considered not only the extreme cases of cancellation and rein-forcement—of destructive and constructive interference—introduced by Young, but also the full range of phase variations of interfering wave trains. He divided each wave into component waves of differing phase which build up indepen-dently, and combined them to determine the intensity of the light. His wave theory was transverse, rather than longitudinal, affording an explanation of the failure of oppositely polarized light beams to interfere. He explained diffraction in terms of these principles; diffraction patterns result from the interference of secondary waves produced at the edge of the diffracting body. Interference be-tween incident light and light reflected by the diffracting body is not involved, so that the behavior of diffracted light does not depend on compositional properties of that body. Fresnel developed an analytic, mathematical theory able to predict positions of fringes and variations in the intensity of light with distance from the borders of geometrical shadows of diffracting bodies of various types.[1]

His predictions accorded well with experimental knowledge, and this success was primarily responsible for his receipt of the 1819 prize awarded by the French Academy for a theory of diffraction. What the academy was looking for, and found, for the first time, in Fresnel's work, was the derivation of properties of diffracted light—the positions and intensities of diffraction bands—from a mathe-matical theory, rather than their determination by experiment. Fresnel's analysis of diffraction was completely general, and owed nothing to experimental knowl-edge of diffraction patterns.[2] This permitted the independent comparison of opti-cal theory with experiment. The experiments that Fresnel himself performed, ex-hibiting a variety of diffraction patterns in greater detail, were directed not at increasing empirical information as such, but at testing his theory. So impressive was this achievement that Fresnel was awarded the prize despite the general inde-pendent opposition of the judges to the wave approach and its associated method of hypothesis.[3]

But the confirmation of Fresnel's theory by observations of the positions and intensities of diffraction bands is not novel confirmation. The use of Young's law of interference to explain diffraction also predicted these bands. And it delivered

1. G. N. Cantor, in *Optics after Newton* (Manchester, U.K.: Manchester University Press, 1983), pp. 156–157, reproduces Fresnel's graph of the variation of intensity of light outside and within the geometric shadow of a diffracting body.

2. "This general principle must apply to all particular cases," says Fresnel, emphasizing the inde-pendence of his analysis from observational data; "Memoir on the Diffraction of Light", in Henry Crew, ed., *The Wave Theory of Light* (New York: American Book Co., 1900), p. 108.

3. Recall the discussion in chapter 2 of methodological controversy associated with the wave theory.

first positions of minimum light intensity, within and without the shadow cast by a diffracting body, with an accuracy close to Fresnel's. Indeed, some of Fresnel's experimental work was devoted to obtaining these positions with an accuracy capable of deciding between the two predictions.[4] By contrast, as suggested in chapter 2, there was no basis, independent of Fresnel's theory, for anticipating the phenomenon of a bright spot appearing at the center of the circular shadow produced in spherical diffraction. Poisson pointed out this implication of the theory, by a geometric analogy to Newton's rings, expecting thereby to discredit it. Had the bright spot not been observed, in the subsequent experiment of Arago, it is doubtful that the prize would have been awarded to Fresnel despite his successes, given the theoretical and methodological predispositions of the judges.

So while the view, common in the philosophy of science, that this prediction won the day for Fresnel and was pivotal in converting optical theorists to the wave theory is simplistic,[5] this prediction is not without historical importance. Little was made of the success of the prediction, but much would have been made of its failure. That the prediction received scant attention in the citation for the award is understandable without supposing that its novelty was immaterial in the evaluation of Fresnel's theory. It is sufficient to note, in addition to the allegiances of the judges, the overriding interest in obtaining a mathematical theory capable of explaining and predicting diffraction patterns already known.

From a philosophical perspective, the significance of the prediction is great. A theory whose provenance owed nothing to known facts about diffraction, a theory intended to yield quantitative predictions for such cases as single slits and straight edges, was the basis for the discovery of a qualitatively distinct effect. A minimally adequate reconstruction of Fresnel's reasoning would cite, as rationale for his physical assumptions, optical effects and properties that bear favorably on the wave theory generally: the analogy to the transmission of sound; the constant velocity of light, and the independence of its velocity from compositional properties of the source, which would be expected to affect light corpuscles; the production of alternating light and dark bands in interference; the success of the Huygens principle in accounting for refraction. Diffraction effects would not be cited at all. They arise as applications, once the theory has been developed. Not only the prediction of the bright spot for spherical diffraction, but also diffraction phenomena in general, satisfy the independence condition for novelty.

But because of Young's work and the extension of particle and ray methods to known diffraction effects, only the bright spot satisfies the uniqueness condition with respect to Fresnel's theory. No particle or ray theory, including Young's interference theory of diffraction, gave any reason to expect this effect, in any magnitude. Its bearing on the evaluation of Fresnel's theory is therefore markedly different from that of the quantitative agreement he obtained for other types of diffraction. This difference is a consequence of its satisfying the uniqueness condition.

4. Cantor, *Optics after Newton*, p. 155.
5. See, e.g., Giere, "Testing Scientific Hypotheses."

Special Relativity

According to my analysis, determining that a result is novel for a theory requires information about how the theory was developed. It is against the logical structure of that development that the result's novelty is judged. If novel status is to make an epistemic difference in confirmation, then my analysis disputes the traditional dichotomy between development (misleadingly called "discovery") and justification.[6] What matters, specifically, is the role of experimental results in the development of a theory for which a given result is alleged to be novel. Thus, in assessing the novelty of the central bright spot in spherical diffraction for Fresnel's theory, it was crucial to identify the optical effects on which the initial plausibility of the physical assumptions that Fresnel used to explain diffraction depended. It was necessary to show that the bright spot need not have been included among these effects; that the plausibility they conferred did not depend on including this result. This was relatively easy; as the bright spot was in fact unknown and unanticipated prior to the Fresnel theory, it did not take much historical analysis to make the case for its independence, in the sense of my condition, from the theory's provenance. It was necessary only to see that the optical effects on which the theory did depend do not generalize in a way that encompasses the bright spot as an instance.

The question of novel status with respect to Einstein's special theory of relativity is more complicated, for it arises in connection with a result that had been known long before the theory was developed and that had already been incorporated into prevailing theory: the null result of the Michelson-Morley experiment. It is not vital here to address the vexed question of *how well* the result had been incorporated. I contend that it was not well enough to violate the uniqueness condition and, thereby, to disqualify the result as novel for relativity.[7] But whether this is correct proves incidental, as it turns out that the result is disqualified in any case, on grounds of independence. As this outcome makes no appeal to the result's antecedence, nor to the possibility of using it to generate the theory, it serves particularly well to illustrate distinguishing features of the analysis. On the other hand, a result that I do hold novel for relativity, the conversion of matter and energy, while incontestably independent of the theory's genesis, raises questions of alternative anticipation and so serves as a good application of the uniqueness condition.

Special relativity was developed from two hypotheses: the independence of the velocity of light and the relativity principle. The first denies that the velocity of light is affected by the velocity of the emitting body. The second, in one form, asserts the equivalence of all inertial frames for the formulation of all natural laws; in another from, it asserts the impossibility, in principle, of any determination of absolute velocity, whether from mechanical or electromagnetic phenomena. The application of these hypotheses to Maxwell's equations for electricity and magne-

---

6. See J. Leplin, "The Bearing of Discovery on Justification."
7. See J. Leplin, "The Concept of an *Ad Hoc* Hypothesis."

tism produces the relativistic effects on the measures of space and time, in the form of a new system of transformations for inertial reference frames. I am concerned with the role of empirical results in the reasoning by which these hypotheses were motivated and applied.

The route to the relativity principle generalizes the more limited relativity that holds within Newtonian mechanics to electromagnetism. The grounds for the generalization are certain suggestive experimental results in electromagnetism and optics, which Einstein cited in the opening pages of his seminal 1905 relativity paper, "On the Electrodynamics of Moving Bodies," [8] and some thought experiments—one about an asymmetry in the prevailing explanation of electromagnetic induction from that paper, and one about motion at the speed of light, reported in Einstein's *Autobiographical Notes*.[9] It is conceivable that the thought experiments alone were enough of a motivation for Einstein personally, that the actual experiments he specifically cites were intended pedagogically. Be that as it may, the actual experiments were crucial to the case that had to be made for generalizing relativity to electromagnetism, and must figure in any reconstruction of the reasoning that produced Einstein's theory. Without them, there is nothing but an intuitive preference for a particular form of simplicity to stand against the successful body of electromagnetic theory that denied the principle.

Einstein says that in the thought experiment about motion at the speed of light "the germ of the special relativity theory is already contained." But this does not quite mean that the experiment reveals the relativity principle; rather it seems to presuppose the principle. Einstein says that nothing in experience (nor, of course, in Maxwell's equations) corresponds to the idea of being at rest with respect to a beam of light. Described from a position of rest, the beam would behave in a distinctive way. This would identify one's speed as that of light; it would permit an inference from relative motion to absolute motion. If, as the relativity principle requires, absolute motion cannot be determined, then motion at the speed of light must be impossible and the laws of nature must be such as to preclude it. In Einstein's interpretation of the experiment, the relativity principle is used to infer a general restriction on the form of natural law. It is not really explained why the relativity principle is to be believed in the first place, except to note that all experience conforms to it.

The net impression is that the importance of this thought experiment was to crystalize in Einstein's thinking the role of the relativity principle in imposing constraints on permissible laws, which were needed because it had proven too difficult to construct laws directly from experiments. The relativity principle is not so much being defended here as being recognized and clarified through application.

The point of the other thought experiment is different. Maxwell's equations give us different explanations of the same observed phenomenon, induced elec-

8. In A. Einstein et al., *The Principle of Relativity* (New York: Dover, 1952).

9. In P. A. Schilpp, ed., *Albert Einstein: Philosopher-Scientist* (New York: Harper Torchbooks, 1959).

tric current, depending on which of two objects in relative motion, a magnet or a conductor, we take to be the one actually moving. There is a conceptual asymmetry that does not correspond to any asymmetry in the phenomena. Einstein reasons that if relative motions are all that the observations depend on, then only relative motions should figure in their explanation. There should thus be but one explanation wherever relative motions are the same, notwithstanding the possibility of different absolute descriptions.

This is a kind of thought experiment, in that it asks us to recognize that differently described actual experimental situations are really the same. This recognition does offer some defense of the relativity principle on methodological grounds. In insisting that the concepts with which we understand observations be no more complex ontologically than the observations require, Einstein denies a role for absolute motion in our theories. If we do not see it, or see anything that requires it, then we should not use it conceptually. But eliminating absolute motion from theory is tantamount to the relativity principle in the form of a constraint on natural laws.

Such methodological reasoning is characteristic of Einstein. It enabled him to reach conclusions from experiment that others familiar with the same experiments did not entertain. The experiments he uses to motivate the relativity principle are examples. He is not specific, in the 1905 paper, as to which experiments relativity is supposed to be founded on, referring only to the failure of attempts to measure the earth's motion relative to the ether, and to first-order results showing that electromagnetic laws and mechanical laws hold in the same reference frames. Other sources identify astronomical aberration and Armand Fizeau's measurement of the speed of light in water as the important results. The Michelson-Morley null result, a result of the second order, is explicitly denied to have been influential, and it is unclear whether Einstein even knew of it (although he must have seen papers of Lorentz that mention it).[10] Analyses of novelty that drop the temporal condition typically endorse the novelty of this result, although pedagogical reconstructions of the provenance of relativity theory, including Einstein's own, commonly use this result to represent the theory's empirical basis. Of course, a number of nineteenth-century experiments could be cited to support the other hypothesis, of the independence of the velocity of light—for example, the facts that binary star systems obey Newtonian gravitation and that light passing between grooves of a rotating wheel is unaffected by the velocity of rotation.

Astronomical aberration has to do with the effect on the apparent position of a celestial light source of the motion of a telescope while the light is passing through it. Explaining aberration was an important problem in optics because of limitations on the accuracy of a raylike model for light. It was not this problem that concerned Einstein, but the fact that the amount of aberration—or the amount by which the telescope's orientation must be changed to correct for it—

10. See Einstein's interviews with R. S. Shankland, "Conversations with Einstein," *American Journal of Physics*, 31 (1963):47–57.

seemed to be independent of the earth's motion relative to the ether. It was not thought really to be independent; rather, the failure to detect any dependence was attributed to the fact that the effect was calculated to be of the second order, and so too small to be observed. The effect is a function of the square of the ratio of the earth's velocity to that of light, v/c. To quantities of the first order of this ratio, no effect is predicted and none is observed.

To understand aberration, with Einstein, as an empirical basis for relativity, a very different and much simpler interpretation is required: No effect is detected and, therefore, none is to be assumed, to *any* order of accuracy. As the earth's motion relative to the ether has no bearing on what is observed, no such motion is to be assumed. This view conforms to Einstein's rejection of an asymmetric explanation of electromagnetic induction, where the observed phenomena, depending only on relative motion, are symmetric. Only the present case is more radical, in that there was a theoretical basis for predicting an effect, and reasons to expect it to become detectable. The strong appeal to parsimony here is reminiscent of Newton's rules of reasoning: Hypotheses should introduce no complications not strictly required by the phenomena. Fidelity to this methodological precept explains why the failure to detect ether-relative motion even at the second order, as in the Michelson-Morley experiment, is to be regarded as confirmation of what one already expects, rather than as a result significant in forming one's expectations. And Einstein was quite clear that the Michelson-Morley result, whenever he first learned of it, occasioned no surprise and made no difference in his thinking.

In citing astronomical aberration as an empirical basis for relativity, Einstein might have intended also to encompass nineteenth-century attempts to measure absolute motion astronomically. These experiments are supposed to yield first-order effects of absolute motion, and construing them as support for relativity would fit his methodological precepts. Olaus Romer deduced the velocity of light from delays in the eclipses of Jupiter's satellites due to variations in the distance light must traverse as the satellites orbit Jupiter. On this method, the ether-relative motion of the solar system would affect the eclipse times by about one second over a six-month interval. That this quantity was technologically undetectable would not have deterred Einstein from counting it as another instance in which absolute motions required by theory are unrealized. Another example is Arago's failure to detect a shift in the focal point of a telescope over a six-month interval, as a star is observed with the telescope directed parallel to the earth's orbit. This shift is only of the second-order if one assumes Fresnel's version of ether theory, which was supported by results obtained by Fizeau and Airy. It should not, then, have been detected. Yet to Einstein it evidently signaled, once again, the nonexistence of absolute motions required by theory.

Although Fizeau's experiment supports Fresnel's theory, Einstein cites it specifically as support for the relativity principle. This interpretation assumes that the contraction effect for moving bodies reduces the length of the light path in moving water, for the relativistic analysis of Fizeau's experiment uses the relativistic theorem for the addition of velocities. This seems to be getting ahead of the game. What Einstein may have intended, however, was to call attention to the fact that the amount of convection measured in the experiment was less than expected on ether theory, and so could be considered another null result.[11]

The upshot of this analysis is to reinterpret a number of first-order experiments, well understood in terms of nineteenth-century ether theory, in conformity with a general methodological precept that dismisses quantitative constraints on their significance. Einstein notes that all known electromagnetic, including optical, phenomena reveal no effect of absolute motion to the first order of v/c. He infers that all such phenomena obey the relativity principle unqualifiedly—to all orders of v/c. The inference is rendered deductive by assuming the methodological premise that no quantitative constraint is to be placed on the generalization of observations, unless required by experience.

Einstein then combines his generalization with the light principle, which is independently generalized from experiments. He adds, on the strength of the empirical success of Newtonian mechanics, that the principle of relativity holds for all mechanical phenomena, and concludes that all physical phenomena are relativistic to all orders of v/c. From this conclusion he deduces a general constraint on the form of physical laws, whether of mechanics or electromagnetism; they are to be relativistically invariant. The identification of the reference frames with respect to which this invariance holds is given by Newtonian mechanics; they are empirically fixed as the frames with respect to which Newton's laws of motion appear to hold, at least approximately—the inertial frames. The independence of the velocity of light may now be cited as a physical law. Consequently, light has the same velocity in all inertial frames.

The kinematics of special relativity now follow deductively. The Michelson-Morley result is a special case, on the assumption that, to good approximation, the earth, relative to which the Michelson interferometer is at rest, is an inertial system. The experiment divides a light beam into two pencils made to traverse mutually perpendicular paths of equal length, measured in the earth frame. Since the paths differ in their orientation relative to the earth's presumed direction of motion through the ether, the speed of light must differ along them if it is to be constant relative to the ether. This difference would be expected to produce a difference of travel times for light along the two paths, with resulting interference occurring when the pencils recombine. There will be some interference in any case, even discounting the difference in speeds, due to incidental features of the apparatus and of the mechanism for producing the light. But given the difference in speeds along the paths, the interference fringes produced will change position as the apparatus is rotated, changing the paths' orientation with respect to the earth's motion. This effect, though of the second order, would be readily observable. The result is null; no such change of position is observed. But this is exactly to be expected according to relativity theory, for the earth has no absolute, or ether-relative motion. The velocity of light is the same in all (inertial) frames,

---

11. Additional analysis of the use of experiments to motivate the relativity principle is given in J. Leplin "The Role of Experiment in Theory Construction," *International Studies in Philosophy of Science*, 2 (1987):72–83.

and the condition of motion relative to the ether—which now, for want of any role, drops out of the theory—can make no difference.[12]

On this reconstruction of the reasoning that led to special relativity, the Michelson-Morley null result is simply an instance of an empirical generalization that serves as a step in deducing the theory. This step is not quite a premise; it is deduced from an empirical premise and a methodological premise. The result is not itself part of the inductive basis of the empirical premise, which is limited to first-order results. But as an instance of a generalization whose role in deducing the theory is crucial, the result cannot qualify as novel for the theory. That Einstein did not know of the result (we may suppose), was not guided by it, did not use it, did not set out to explain or predict it—none of this is material. For we cannot omit from the reconstruction, without ruining its validity, a step that implies it. The result is logically an ingredient in the foundations of the theory, even if it is psychologically and historically absent from them. The theory therefore depends on the result, in the sense of my independence condition for novelty. Specifically, a qualitative generalization of the result, denying that absolute or ether-relative motion affects optical phenomena, is ineliminable from a minimally adequate reconstruction of the reasoning that produced the theory.

The situation is otherwise for the convertibility of mass and energy. This result, deduced from the theory in 1905,[13] and established empirically by particle accelerators in the 1930s—and, more generally, by the phenomenon of radioactivity— has no basis in the electromagnetic laws to which the kinematics of relativity were applied to produce a full, electrodynamic theory. It bears some superficial, formal resemblance to an earlier result of Lorentz, and, for this reason, might be suspected of violating the uniqueness condition. But the Lorentz result has a very different, and much more limited, significance.

The fact that energy possesses an inertial resistance may be deduced in relativity theory from the variation of mass with velocity. This result is obtained by applying the transformations required for the relativistic invariance of Maxwell's equations—the Lorentz transformations—to the mechanical law of conservation of momentum. For this law to obey the relativity principle, when the Lorentz transformations are used to relate inertial frames, introduces a dependence of the mass of a body on the choice of inertial frame. This amounts to an increase in mass with increasing velocity, from the perspective of any given frame. If a body with mass m_r at rest moves with velocity v, its mass becomes $m_r / \sqrt{1 - v^2/c^2}$.

More precisely, relativity theory replaces the classical expressions for momentum and kinetic energy with analogues that remain invariant under the Lorentz

---

12. It might seem that the null result could be reached more directly—simply by applying the relativity principle to optics. If light propagates isotropically in any one frame, it must so propagate in all frames. And of course, light would be expected to propagate isotropically in the ether frame. But one cannot consistently single out one frame as the ether frame and also uphold the relativity principle in optics. And to appeal to experimental indications of isotropic propagation is insufficient without introducing the generalizations that Einstein was prepared to make.

13. Albert Einstein, "Does the Inertia of a Body depend upon its Energy Content?", in *The Principle of Relativity*.

transformations. To recover the classical law of momentum conservation for the new expressions in the case of inelastic collisions, it is then necessary as well to extend the classical law of kinetic energy conservation to inelastic collisions. This, however, associates an energy of mc^2 with a body of rest mass m. In this line of reasoning it is not necessary to speak of mass as a function of velocity, and the classical conception of mass as quantity of matter can be retained. However, mass understood as inertial resistance does increase with velocity, for inertial resistance increases with kinetic energy.

The same functional dependence of mass on velocity had been obtained by Lorentz for the electron, from the fact that electric charge increases inertial resistance. This is anticipated by Maxwell's theory. A moving charge constitutes a current which generates a magnetic field. Since a field carries inertia, a moving charge, with its associated magnetic field, carries greater inertia than a charge at rest. In the Lorentz theory of electrons, this results from the application to moving electrons of the contraction effect for moving bodies—which Lorentz hypothesized to take account of the Michelson-Morley null result—in combination with Lorentz's analysis of the force exerted on an electron by the electric field that it itself generates.[14] The electron's mass, or inertial resistance, is supposed to be electromagnetic in origin, arising entirely from the electron's charge. The velocity on which the electron's mass depends is that which produces its contraction, and must be an ether-relative velocity.

By contrast, Einstein's result is completely general, and unlike Lorentz's, assigns an energy to a mass at rest. It depends on no assumptions about the ether or electricity or the electrical basis of inertia, but only on the application of the relativity principle to conservation of energy and momentum. The velocity on which mass depends may be taken relative to any inertial system. Experimental knowledge prior to 1905 of the effect of velocity on electronic mass is no confirmation of Einstein's far more general result, and is irrelevant to the novel status of mass-energy equivalence for relativity theory. One could not generalize Lorentz's result into Einstein's, even if there were some motive to do so, without contradicting other tenets of the Lorentz theory—the asymmetric or nonreciprocal nature of the Lorentz's ether-theoretic transformations, for example. On the contrary, the generalization suggested within the Lorentz theory is that all mass is electromagnetic in origin, and that, ultimately, mechanical laws are to be reduced to electromagnetic laws. This would mean that the laws of mechanics do not obey the relativity principle, after all, but only appear to do so where quantities of the first order alone are considered. Such a development is completely contrary to Einstein's theory, which extends the mechanical principle of relativity to electromagnetism. Einstein's theory yields the variation of mass with energy as a property of mass understood as inertial resistance as such, without any restriction to an electromagnetic origin.

Thus, mass-energy equivalence, manifested in the conversion of the kinetic energy of particles into the mass of new particles in accelerator experiments, pro-

---

14. H. A. Lorentz, *The Theory of Electrons* (New York: G. E. Stechert, 1915), pp. 38–39.

vides a contrasting case to the Michelson-Morley result. The former meets both conditions for novelty. Its mechanical basis is independent of the provenance of relativity, which takes from mechanics only its consistency with the relativity principle. The apparent anticipation of mass-energy equivalence by the dependence of the electron's mass on velocity in earlier electromagnetic theory proves illusory.

The Expansion of the Universe

I have addressed the question of novelty for some classic tests of general relativity, but there is another test of the theory that is not considered classic and is not usually cited as part of the theory's empirical support. This is the prediction of the instability of the universe as a whole. The status of this prediction has been confused by Einstein's introduction of the "cosmological constant," a term in the field equations of general relativity that is not fixed by the theory, but that may be adjusted to bring the theory into conformity with observations. On the assumption that the universe, as a whole, is stable, Einstein initially assigned this term a nonzero value to offset an instability that the equations otherwise imply. But the universe turns out not to be stable. Some thirteen years after Einstein published the theory, Edwin Hubble's observations of the redshift of distant galaxies indicated that they are receding from us with velocities proportional to their distance, indicating an overall expansion of the universe. Resetting the cosmological constant to zero—or eliminating it from the theory—brings the theory into conformity with this universal expansion, and it is common to assert that had Einstein left his equations unadulterated in the first place, he would have been credited with a major discovery that was subsequently confirmed empirically. By unimaginatively presupposing the stability of the universe when his own theory, taken at face value, suggested otherwise, he put himself in the unfortunate position of responding to observations rather than anticipating them.

On my analysis of novelty, however, the difference between response and anticipation need not affect the epistemic import of prediction. And it is something of a confirmation of this feature of the analysis that the usual depiction of Einstein as having missed out on a major discovery does not ring true. Einstein *himself* may be faulted for sticking to the assumption of a stable universe (and he so faulted himself), but what has that got to do with whether *his theory* is to be credited with a predictive success? And if the theory is credited, does that not redound to the credit of its author? A scientist gets no credit for predictions made without theoretical basis; they would be more guesses than predictions, their verification luck rather than merit. Why, then, not grant him credit when he does have a theoretical basis, even if he personally is insufficiently audacious to embrace the consequences? Max Planck is regarded as the father of the quantum theory because he introduced the quantization of radiation in calculating the spectrum of blackbody radiation. But Planck did not endorse the physical implication of his radiation law that energy is quantized. He explicitly disowned the quantization effect, regarding it as a temporary mathematical expedient. Planck is credited because his law provided a theoretical basis for energy quanta. It is theories that we are evaluating when we measure predictive success, not theorists. I imagine that had Einstein responded to Hubble's law by claiming that it vindicated his theory and removed the need to detract from its elegance by adding an extra term without a theoretical basis, rather than by faulting himself, he would have gotten better press. Credit goes to those who claim it. This sorry saga of public relations should not be allowed to decide an epistemic issue.

The expansion of the universe, or, at any rate, its instability, was certainly a successful novel prediction of general relativity. Its novelty is clearer than that of the classic tests. There is no connection between the empirical principles that had a formative role in the development of the theory—the principles of equivalence and general covariance—and the recession of galaxies. And there was no competing theoretical basis for predicting expansion. The basis in general relativity for instability lies simply in the fact that the geometry of space-time in the presence of matter is not Euclidean. The curvature of space-time is enough to support the inference that the universe is nonsymmetric in time, that it will appear structurally different at different times.

The Big Bang

The expansion of the universe implies its evolution from an earlier state of greater density. Its expansion according to the equations of general relativity implies, more specifically, that the universe originated in a state of infinite density, a "singularity" at which the equations break down. This "big bang" scenario carries testable implications.

One is that the density of distant galaxies should appear greater than that of local galaxies, because of the connection between distance and age. More distant galaxies are observed at earlier times, because of the time required for the transmission of light.

Another implication is that radiation produced by the big bang should continue to permeate the universe. Such radiation should be observed equally in all directions (but for minute variations owing to inhomogeneities in the early universe that result in the local nonuniformity of the present universe) and so be readily distinguishable from other radiation sources. Because the early universe is in thermal equilibrium, its radiation must exhibit a distribution of frequencies characteristic of a blackbody, or perfect radiator, which emits all the energy that it absorbs. For a given temperature, such black radiation shows a characteristic spectrum, or distribution of energies. So the spectrum of big bang radiation in the present universe is predictable, if the age of the universe is used to calculate the temperature to which the radiation has cooled. Also required, however, is a detailed theory of the formation of the heavier elements by nuclear fusion. Such a theory was not immediately available, so a quantitative measure of the expected spectrum of residual radiation was delayed.

Both predictions are novel for big bang cosmology. Neither they nor any generalization of them is involved in the provenance of the theory, which comes directly from general relativity, the simplifying assumptions used in solving the theory's equations, and the generalization of Hubble's law to all parts of the universe. There is no independent theoretical basis even for qualitative generalizations of these predictions. This is especially significant in connection with the second prediction, as quantitative estimates of the present temperature of residual radiation varied considerably, both as a result of uncertainty over the age of the universe and as a result of difficulties in filling in gaps in the fusion process. If another theory predicted a nearly homogeneous, universal radiation source that differed only quantitatively from that finally deduced from the big bang theory, then the result would not be novel. The state of collateral knowledge would have afforded too much opportunity to adjust the quantitative accuracy of a competing theory.

Moreover, both predictions have been verified. Distant galaxies not only are more densely distributed, but also, in several respects such as ongoing star formation, they appear younger. The verification of the predictions, particularly the second prediction, was historically decisive in favor of the big bang theory.[15] In this example, the epistemic importance of novelty registered acutely within the scientific community.

Admittedly, this does not always happen. The definitive influence of an empirical result, even if novel, on the fortunes of a theory is unlikely, if only because it takes time to absorb major changes in one's justificatory situation. If so revolutionary a thinker as Einstein, convinced of general relativity, could not infer the instability of the universe without further evidence, it must be expected that considerable reflection about and review of alternatives will be required before scientists embrace a position for which they already have epistemic support. And the epistemic support provided by a particular result does not necessarily deserve to be definitive simply in virtue of the novelty of that result, if only because other novel predictions of the theory might fail. Added to this is the historical problem of estimating the reactions and convictions of earlier thinkers, whose standards and motives might have differed from those that their recorded public discourse inclines us to attribute to them. It is far from clear what was the historical influence of the novel predictive success of Fresnel's wave theory. It cannot be said that the importance of novelty is always felt, nor that we can always determine whether it is felt.

Equally problematic, as the historical requirement for novelty anticipated, is the determination, within a historical context, of novel status itself. To decide this requires assessing the capacities of rival theories to yield empirical results. This in turn demands that we circumscribe the range of auxiliary information secure enough to be assumed in deriving predictions. We must be able to determine both whether the current state of such information enables rival theories to predict the same result, and whether the auxiliaries on which the predictions of earlier theories depended are still viable. The conclusions we need may be difficult to reach, tentative, and controversial. What results a theory can and cannot

15. Stephen G. Brush, "How Cosmology Became a Science," *Scientific American*, 267 (1992):62–70.

account for may be far from clear over a considerable period, and may be settled only by inquiries into auxiliary matters which the question itself prompts. The shifting status of auxiliary information makes its circumscription inexact, and resulting findings of novelty problematic. With historical distance, the reliability of pronouncements as to novelty increases even as that of attributions of its significance to the scientists involved decreases. Especially if a result is temporally new, so that there has not been opportunity to investigate avenues of explanation independent of the theory from which it is first predicted, will novelty be hard to assess—just the opposite of what is commonly supposed.

These problems do not, however, prevent us from attaching epistemic importance to novelty and assessing the deserts of theories accordingly, if only in retrospect. Such assessments can be clear and warranted, even if they do not look that way in historical context. To suppose that absent unanimity of conviction, no objective verdict for theories is possible, is simply a non sequitur. If objectivity were problematic, unanimity would not be the solution anyway; everybody could be wrong. It is the relation of warrant between evidence and theory that matters, not how readily or unequivocally this relation is recognized.

The first prediction of the big bang theory was verified by a group of astronomers at Cambridge University in the early 1960s, by counting radio sources from outside our galaxy. The ages of distant galaxies were estimated from the spectral slopes of their radio emissions. Their numbers, or density, could then be compared at different ages, or distances. The Cambridge astronomers were proponents of a steady-state theory of the universe, which competed with the big bang theory. In the steady-state theory, a kind of stability is achieved for the universe, despite the expansion established by Hubble, by positing a compensating creation of new material in spaces left open by expansion. Thus, the finding of differences in the density of local and distant sources, which indicated the instability expected on the big bang theory, contradicted the expectations of the Cambridge group.

This situation is reminiscent of the Fresnel case, in which particle theorists sought observational evidence against the competing wave theory, only to disconfirm their own view. When results that support a theory are obtained in this way—through experimental or observational research for which the rationale and expectations are opposite to the outcome—it is especially clear that the results are not implicated in the provenance of the theory they support in a way that would defeat their novelty. Of course, they could be so implicated; the attitude of the scientists involved is not definitive. But getting the opposite of what one's theory leads one to expect is reliably indicative of independence, in the sense of the analysis. And at a time when most science-studies disciplines equate the presumed "theory-ladenness" of observation with the impossibility of objectivity, it is therapeutic to note such cases. These are nice cases, in that the epistemic import of the results is evident, even if sometimes unacknowledged for the reasons just given. Such cases give us paradigms of novelty that any philosophical analysis must respect. But they are atypical. Usually, the question of whether a successful prediction is rightly interpreted as testifying to the correctness of the theory that predicts it, as opposed to being explainable in some other way, is harder to answer. This more common difficulty is reflected in the complexity of my analysis. Novelty, on my analysis, typically requires a detailed inquiry into the provenance of theory to assess.

The second prediction of the big bang theory, the prediction of the cosmic background radiation, was verified by Arno Penzias and Robert Wilson of Bell Laboratories. These investigators did not have relevant theoretical expectations, and did not, at first, appreciate the theoretical significance of their inability to eliminate a persistent, microwave interference from a satellite communications receiver. But they reported their problem to some Princeton cosmologists—Robert Dicke and James Peebles—whom they heard had a theory that could explain the microwave signal. This was not quite the big bang theory, involving the origination of the universe in a singularity, but it was close enough. It depicted the universe as undergoing cycles of expansion and contraction, each cycle originating in a concentration of matter of sufficiently high energy to leave a blackbody signature as the universe cooled. This explanation of the observed expansion, and other accounts, short of the big bang, of the initial conditions from which the universe evolved, were ultimately eliminated by the singularity theorems of general relativity. These theorems, published by Stephen Hawking and Roger Penrose in 1970, showed, in essence, that general relativity requires the universe to have begun in a singularity.

Thus, strictly speaking, the cosmic background does not require the full big bang theory, as it has come to be understood. But it does require much of the leading idea, enough, for example, to establish the theory over its steady-state competitor. Similar observations are appropriate with respect to other examples of novelty. It is not Fresnel's particular theory, to the exclusion of any other way of implementing the wave approach to light, that spherical diffraction necessarily supports. To claim the epistemic import of novelty for an empirical result is not, admittedly, to claim correctness for the specific details of the theory for which it is novel, as opposed to variations on the same basic theoretical program. Some changes or resolutions of detail are changes *within* a theory rather than changes *of* theory, the same theory being reidentifiable through them. Any inference to the correctness of a theory from the success of its novel predictions must be qualified accordingly. Also, the absence of a competing theory predicting the same result, to within a qualitative generalization of it, which the novelty of a result requires, is obviously temporally indexed; nothing guarantees that no such theory will be forthcoming. Any claim that a result "requires" a theory must be qualified accordingly. The requirement is relative to a changeable epistemic situation, in which certain results are known and certain theories are available.

These admissions raise the general question of what, exactly, the epistemic import of novelty is supposed to be. What, as I have put it, is the relation of warrant between evidence and theory, when the evidence includes novel results? What can be inferred about theories from their novel predictive success? What is the connection between novelty, as I have analyzed it, and the general philosophical position of realism, which motivated my inquiry into novelty in the first place? With an analysis of novelty on offer, we are prepared to answer these questions.

5

Realism and Novelty

The Burden of Argument

Having motivated, formulated, and applied an analysis of predictive novelty, I turn to the question of its epistemic significance. I wish to show that and how novel success supports theories. As observed in chapter 2, some philosophers deny that theory confirmation depends on novelty, and argue that the novelty of a result can make no difference to the warrant it bestows on the theory that predicts it. I attribute their conclusion to the presupposition of an inadequate conception of novelty. No amount of philosophical effort will defeat the intuition that novelty matters, so long as a better conception is available to sustain it. I have now provided one.

What is the advantage in this? When theory and intuition clash, either can be at fault. Of course, if the theory is a philosophical theory whose subject matter is intuition—how our concepts work—then its ultimate test is getting the relevant intuitions right. To the extent that it departs from them, its vindication requires recourse to deeper intuitions in terms of which the ones it violates can be corrected and improved. By contrast, physical theories may be counterintuitive without prejudice, because their ultimate responsibility is not to how we think, but to facts about the world that ordinary thought can misrepresent.

The theories about science at stake in judging the importance of novelty are not happily classifiable by these extremes. For one thing, it is not clear that philosophical theories need, or should, take intuitions as their subject matter. Traditionally, they do, and contemporary ethics and metaphysics generally uphold this tradition. But increasingly in epistemology, and especially in the philosophy of science, theories have been redirected toward methods of inquiry that can be empirically assessed for their utility in advancing cognitive goals without prejudice as to whether they are believed by practitioners to advance these goals, or whether they fit popular preconceptions as to what knowledge is or how it is acquired. This development abandons philosophy's traditional claim to a priori status in favor of an essential continuity of philosophy with empirical science.

The subject matter of scientific realism is not method, but the theoretical conclusions to which the methods of science lead. How these conclusions relate to the world is neither a purely conceptual issue nor a purely empirical one, but depends on considerations of each type. This complexity has caused great confusion, ranging from Ernest Nagel's dismissal of the issue as "terminological"[1] to Thomas Kuhn's rhetorical admonition that settling it would require access to the world independent of theory to determine whether they match.[2] The resolution I wish to defend requires more than conceptual argument, but less than the elusive standard of unvarnished empirical proof. The contribution I seek from intuition is help in fixing the burden of argument. Specifically, the intuition that novelty counts favorably in theory evaluation is important for what it indicates about scientific practice, and scientific practice is entitled to a presumption of warrant.

That the intuition is strong within science, though relatively uncontroversial, is difficult to prove. Gardner reports that "the history of science provides adequate evidence that the superior supportive force of 'novel' facts has played a significant role in scientific reasoning."[3] This assessment, carefully vague though it is, may be overstated. We have striking examples, like the conversion to atomism of its arch enemy, Ostwald, in the face of the novel success of Einstein's analysis of Brownian motion. But we have no definitive survey of the responses of scientists to successful predictions, measuring the impact of novelty on their conclusions. Such a survey would probably not be feasible, because of the problems, discussed in chapter 4, in determining, in historical context, what results are to count as novel for a new theory. And such historical work as has been directed at the impact of novelty is compromised by the confusion of novelty with temporal newness, and by naïveté as to philosophical subtleties that arise in deciding what is to count as novel.[4] These problems reflect an understandable looseness and variability in uses of the term within science, which it is a purpose of philosophical

---

1. Ernest Nagel, *The Structure of Science* (New York: Harcourt, Brace, and World, 1961), chap. 6.

2. Thomas Kuhn, *The Structure of Scientific Revolutions*, pp. 206–207.

3. "Predicting Novel Facts," p. 2.

4. Identifying novelty with temporal newness, the historian Brush questions its importance because physicists were as impressed by general relativity's ability to yield accurate predictions for Mercury's precession as by its prediction of light deflection ("Prediction and Theory Evaluation: The Case of Light Bending," *Science*, 246 (1989):1124–1129). This may reflect doubts about the reliability of Eddington's results. In any case, Brush's own study of the scientific status of cosmological theories, "How Cosmology Became a Science," which considers cases whose novelty is less controversial philosophically, suggests the opposite conclusion.

analysis to identify and correct. Because the historical determination of what particular scientists have thought or done demands fidelity to a wider range of linguistic practice than is recoverable from a philosophical analysis focused on the epistemic import of a concept's use, the record of practice that historical studies provide is bound to be equivocal. Nevertheless, the expectation that such a survey, were it feasible, would support Gardner's claim is well founded. It will seem the more so if we consider the reasons why novelty should matter.

Special value is placed within science on the explanatory resources of theories. A theory that faces and meets explanatory challenges is valued over one that does not, even if the theories are equally compatible with a common body of relevant evidence. And a novel result, because of its independence in the sense of my condition, represents an impressive explanatory achievement for the theory that predicts it. One way that a theory displays explanatory power is by successfully predicting a result from which its provenance is free. If no other theories predict even qualitative generalizations of the result, then there is an explanatory challenge that the theory uniquely meets. In this situation, if we are not to credit the theory with some measure of truth, then we have *no* way to understand how the theory is able to meet this challenge. The result itself need not be considered mysterious, in that we have a theory to explain it. But this explanatory achievement is itself a mystery. Nothing but the theory's truth explains it.

The claim that empirically successful theories must be at least approximately true was once advanced as the only alternative to acquiescing in mystery. There is no way to understand how science can succeed, it was suggested, if the explanatory mechanisms it postulates are unrepresentative of actual processes in nature.[5] Indiscriminate as to the forms of empirical success that are to betoken truth, this strong realist claim is vulnerable to the fashioning of alternative explanations for various empirical achievements of theories. Much scientific success is readily explainable without imputing truth to the theories that achieve it. The possibilities were canvassed in chapter 1. None of them applies to the special case of *novel* success, as I have analyzed it. Novel success is the exception to the antirealist strategy of proliferating explanations of success; it cannot be explained without crediting the theory that achieves it with some measure of truth.

But explanation is a fundamental goal of science. And this goal cannot be confined to observations, because understanding why theories succeed or fail at predicting observations is crucial to improving theories and so to explaining observations.[6] Since explanation of the fortunes of theory and explanation of the world

---

5. In "What Is Realism?" (Leplin, *Scientific Realism*, p. 141), Hilary Putnam contends that it would be a "miracle" for any (wholly) false theory to predict phenomena successfully.

6. In "The Language of Theories," in Herbert Feigl and Grover Maxwell, eds., *Current Issues in the Philosophy of Science* (New York: Holt, Rinehart, and Winston, 1961), Wilfrid Sellars proposed a view of theoretical explanation on which the proper objects of explanation are neither individual observations nor their lawlike patterns. Rather, theories explain why observable events satisfy general laws to the extent that they do. There is already, in this proposal, a recognition that for scientific knowledge to grow, it is not enough to explain the phenomena directly; the degree of accuracy of our generalizations about the phenomena must also be explained. The progressiveness of theory change

are axiologically interdependent, theories may address other theories as well as the world directly.[7] Thus the explanatory goal of science extends to the success of theories. And if we cannot explain the novel success of theories without supposing them to contain truth, this goal will require a realist interpretation of theories that successfully predict novel results. That is, the explanatory mechanisms of such theories will be taken to represent, at least to some extent, the natural processes actually responsible for what we observe. But explaining novel results realistically means giving them a special probative weight that is not attached to results whose successful prediction by unrepresentative theories poses no mystery. In short, the explanatory ends of science invest novelty with special epistemic significance. They make it plausible to suppose that within scientific practice, the successful prediction of novel results credits theories more than general conformity with empirical evidence.

But to grant this much is immediately to place the argumentative burden on those who would deny novelty its epistemic significance. For how are methodological prescriptions as to what matters in theory evaluation to be assessed, if not by reference to the values that govern practice? A methodology that favors theories only in the measure that they conform to evidence, and does not recognize the epistemic relevance of novelty, is as deficient as an ethical theory that favors conduct only in the measure that it conforms to justice, and does not recognize the moral relevance of altruism. It is, at first blush, a condition of adequacy for a philosophical theory that it respect the relevant, independently motivated precepts. A theory that fails this test, although it could be correct, is deficient in the absence of some additional argument, beyond that from which its specific content issues, to undermine those precepts or relegate them to a different domain of inquiry. The alternative is to assess philosophical theories in an aprioristic vacuum devoid of data—to suppose that, unlike science, philosophy stands or falls on the basis of internal, logical criteria alone. The relevant precepts in the present case comprehend not merely the intuition that novelty matters, but, more im-

---

cannot be understood simply in terms of limiting cases relations, because, among other problems, superseded laws are not just expanded upon but *corrected*. Understanding why they were correct to the extent that they were *and no further* gives us an objective basis for regarding the transition to new laws as epistemically progressive. Without such understanding, we have incompatible systems of explanation that resist neutral comparison. With it, we have explanatory relations among successive stages of science that, in the absence of deductive relations, can underwrite a diagnosis of progressive change.

The complication is that it is not incumbent on theories, as such, to explain the degree of success of their predecessors. This is too strong because theory change frequently reconceptualizes the domains of evidence and information to be predicted and explained, relegating some tasks to other parts of science. The appropriate requirement for progress is, rather, that from the perspective of science broadly speaking, the successes and failures of rejected theories be explainable. For a more detailed treatment of the connection of progress with realism, see J. Leplin, "Truth and Scientific Progress," in Leplin, *Scientific Realism*.

7. A "theory of principle," in Einstein's conception, is a highly general theory that lays down certain constraints that more specialized theories must satisfy. Theories of principle may be formulated explicitly as theories about theories.

portantly, the fact that because of its grounding in the goals of science, this intuition has been implemented in evaluative practice.

The view that philosophical theories of science are to be judged against scientific practice, much as science is judged against natural phenomena, is a form of naturalism in epistemology. A preliminary appeal to naturalism was made in chapter 3. The idea is that epistemological theories actually *are* scientific theories, albeit of a highly abstract and general kind. Just as scientific theories are responsible to the workings of the natural world, a theory about the assessment of scientific theories is a theory about evaluative practice in science, and is responsible to this practice. According to naturalism, there is nothing else for it to be about, for there is no source of facts about how theories ought to be evaluated apart from scientific experience.

Naturalism is not obliged to place scientific practice above critical scrutiny, any more than scientific practice requires that observational data be taken at face value. Science does not abandon objectivity in treating observations as variously interpretable and potentially misleading. That practice is ill conceived or unavailing to the goals of science are possible conclusions of naturalistic inquiry. Naturalism only requires that such conclusions be empirically based, rather than preordained by an antecedent, a priori commitment to a particular theory of method. Perhaps the best argument for naturalism is the insufficiency of aprioristic reasoning to identify uniuely any source of facts about how theories ought to be evaluated, or to decide among potential candidates in a nonarbitrary way.

In the spirit of naturalism, it is reasonable to expect philosophical theories of science to respect the probative weight of novel success. To the extent that they do not, it is incumbent on them to show cause why novelty ought not to matter, and to account for the thereby misguided but otherwise salubrious practice of placing special epistemic value on the sort of explanatory achievement that successful novel prediction represents. It is the argument of this chapter that realism meets this standard, and of this chapter and the next, that rival epistemologies of science fail it. (The limits of the argument are faced in chapter 7.) The ability to rationalize the practice of according epistemic significance to novel results gives realism an advantage that its rivals cannot overcome. Only by interpreting a theory realistically do we explain its achievement of novel success. Rival interpretations bear the onus of explaining why their deficiency in this regard need be tolerated.

Minimal Epistemic Realism

Recall from chapter 1 the thesis that there are possible empirical conditions that would warrant attributing some measure of truth to theories—not merely to their observable consequences, but to theories themselves. This is minimal epistemic realism (MER), a relatively weak form of scientific realism that need not endorse any actual theory. Compatibly with MER, truth may not be attributed to theories, either because the relevant empirical conditions are unrealized, or because the attribution of truth, though warranted, is withheld. It may be withheld even if the warrant for it is appreciated, because the warrant for another course may operate instead. Perhaps an incompatible truth attribution is also warranted; perhaps there is reason to decline to think realistically at all. In general, that an action is warranted does not imply that it is *uniquely* warranted. There may be incompatible routes to a common goal, as well as incompatible goals that cannot be commonly served. The question of whether it is irrational *not* to believe theories does not collapse into the question of whether it is rational to believe them. My quarrel is with those epistemologies that violate MER by denying the possibility of warranted attributions of truth to theories, not with the skeptical option as such.

The case for MER is "explanationist," in that it is based on the requirements for explaining novel predictive success. But we will find that it does not suffer the liabilities that attach to explanationist reasoning generically, in that it carries defeasible empirical consequences independent of the phenomena to be explained. MER itself does not carry such consequences, but MER itself is not proffered as an explanation of anything. Rather, realism, in the sense of a realist interpretation of theory, is proffered as an explanation of novel success, and the fact that realism explains novel success, while carrying further, defeasible consequences, functions as a premise in an argument for MER.

A realist interpretation of a theory takes the explanatory mechanisms postulated by the theory to be representative of actual processes in nature that produce the observationally accessible effects whereby the theory is tested. A realist interpretation attributes some measure of truth to the theory, where truth is understood as accuracy of representation.

As accuracy comes in degrees, it is natural to speak of "partial" or "approximate" truth, or of truth in "some measure." Such qualifications are appropriate because there is no reason to suppose that complete or unimprovable accuracy of representation is required for the explanatory or predictive adequacy of the mechanisms a theory postulates. There is no reason, in general, to assume that the effects that test a theory are capable of discriminating partial from maximal accuracy of its representations, although they may be sensitive to differences in degree of accuracy.

Moreover, complete accuracy is not, in general, a clear conception for representations. Objects are represented in certain respects for certain purposes. For different purposes, different features are salient. Other features may be depicted only roughly or suppressed entirely, producing idealizations. Representations may deliberately include features that do not, or even cannot, apply to what is represented, so that other aspects of more pragmatic importance are represented more conveniently. This is the pattern in analogue models, for example. Drawings or photographs of the same object may be compared as to accuracy of representation, without there being any standard as to what would qualify as a completely accurate representation. The standard for comparison of representations in such cases is the object represented, not some ideal representation of it. In the case of theoretical representations of unobserved objects or processes, comparisons must obviously be indirect, but the moral is the same. Unimprovable accuracy is an elusive concept at best.

It may be questioned whether a sense of "degrees of representational accuracy" based on visual representation can be imported to the context of theories. If this sense depends on independent access to the thing represented, it may not, in principle, be available in the theoretical context. But surely our understanding of degrees of representational accuracy does not depend on there *actually being* independent access to the thing represented, for we extrapolate from cases in which there is such access to cases in which the thing represented is unavailable for direct comparison with our representations—is too large, too small, or inaccessible. We might say of such cases that such comparison is still possible *in principle*, but it is unclear why this condition should mark the boundary of our powers of extrapolation. What makes something visually accessible in principle need have nothing to do with what makes meaningful the application to it of descriptive terms. The *procedures* by which the thing represented would be compared with its representation to test the accuracy of the representation are connected with how such accuracy is to be understood. But there is no reason to suppose that the connection is such that where such procedures break down, the concept of representational accuracy is correspondingly lost. We need such procedures to understand accuracy to begin with, in unproblematic visual cases. But the understanding thereby acquired is subject to extrapolation. Only an implausibly rigid verificationist doctrine of meaning, actually *identifying* meaning with procedures for testing a term's application, could make testability the limit of comprehension. On such a doctrine, one could not understand ordinary descriptions of situations hypothesized to preclude observers.

To interpret a theory realistically is only to suppose that its explanatory mechanisms capture some of the features of natural processes well enough not to be misleading as to how the effects these mechanisms explain are actually produced. A realist interpretation claims that the theory reveals some significant truth about real processes, where "significance" is relevance to our explanatory ends, and "some" is a measure proportionate to those ends. The truth in which realism trades is not the all-or-nothing, bivalent truth that attaches definitively to discrete propositions one by one. The interpretation of qualified attributions of truth, and the grounds for judging the extent to which truth is achieved, will be revisited in the last two sections of this chapter.

The Explanatory Poverty of Instrumentalism

I claim that novel success warrants a realist interpretation of theory. My strategy for making good on this claim has been to analyze novelty in such a way that realism is required to explain its achievement. This is no argument by stipulation. Novelty thus analyzed is clearly achievable, which is all that MER requires, and, as shown in chapter 4, is sometimes achieved. If theories are interpreted realistically, this is to be expected. Insofar as a theory has correctly identified and described the mechanisms actually responsible for what we have observed, it becomes reasonable to expect anticipations of further observations deduced from the theory to be realized. If what the theory tells us produces the effects that it is intended to explain really produces them, then further effects that the theory also

requires should really be produced. We thus expect a successful theory to continue to be successful when pressed beyond the phenomena involved in its development, *provided that we interpret the theory realistically*.

By contrast, if we do not interpret theories realistically, then we have no basis
for expectations of their continuing success. Nonrealist interpretations differ, but
to focus on their common inability to ground such expectations, let us consider
them generically under the heading of "instrumentalism." Originally, instrumentalism was the view that theories are instruments for systematizing the observed
phenomena in pragmatically useful ways. Without theories, we have just random
lists of the phenomena. With them, we have more easily recordable general principles from which the phenomena can be recovered systematically. Thus, representing the sky as a celestial sphere, and the objects in it as embedded in a
sequence of spheres undergoing regular relative motions, vastly improves on the
raw data of astronomical tables and charts. The pragmatic utility of such representations does not depend (except, perhaps, psychologically) on their being treated
realistically. And by the time of Copernicus, it was common *not* to treat them
realistically, in part because of conceptual problems that arose in doing so.[8]

Contemporary instrumentalism allows that theories are not merely conceptual
devices in the service of pragmatic ends, but correct or incorrect representations
of unobservable entities and processes. It allows theoretical pronouncements to
be read literally, largely out of despair at finding an acceptable alternative way to
read them. But it denies that the *representativeness* of theories, the correctness of
their representations, is assessable on the basis of the evidence in principle available to us. Theories must still be judged by instrumentalist standards, even if they
are not themselves to be interpreted as instruments. Thus, the question of truth,
although, in principle, applicable to theories, is held to be irrelevant to the ends
of theorizing and to the evaluation of theories. All that matters is the conformity
of the observable consequences of theories to experience. Truth is a concern only
for the observational portions of theories.

Neither sort of instrumentalism gives us any reason to expect successful theories to be successfully extendable to novel effects. If success does not betoken
truth, then it is not reasonably projectable. Why should a predictive mechanism
that has worked so far continue to work, if there is no basis for understanding *why*
it has worked so far? From an instrumentalist perspective, it is no better than
mere coincidence that it has worked at all. We cannot consistently regard a theory's success as coincidence or chance, and, at the same time, regard it as more
likely than not to be extendable. A theory that has *failed* so far is just as likely to
get further phenomena right as one that has succeeded, if success itself is but
happenstance.

It might be thought that this limitation of instrumentalism applies only to the
extension of theories to phenomena of a kind different from those by which they

---

8. The Ptolemaic system of astronomy lost its physical coherence due to the complications that
fidelity to new observations imposed. But pragmatically, it was as useful as the new Copernican system,
and might have prevailed much longer if instrumental utility were the sole consideration.

have been tested. Continuing success with phenomena of the same kind might be expected on instrumentalist grounds as a simple matter of induction. Thus, once a theory has established its empirical credentials with respect to a certain domain of phenomena, it can be judged reliable, in general, with respect to purposes circumscribed by that domain. So if the successful application of Newton's law of gravitation to the data available to Kepler does not make it likely that newly discovered planets will display Newtonian orbits, at least it underwrites a continuing reliance on Newton's law in connection with known planets. If the satisfaction of the Galilean principle of equivalence by falling bodies is no basis for predicting free fall on the Moon, at least the principle may continue to be applied with confidence on earth.

In these cases, the fact that the further application proved successful was accorded greater confirmatory weight than continuing success with phenomena already predicted, and this provides a natural basis for distinguishing kinds of phenomena. A more rigorous criterion for the individuation of kinds is supplied by the standard of novelty. Thus, it might be thought that the instrumentalist is entitled to pronounce Fresnel's theory a reliable indicator of straight line diffraction patterns, but not entitled to infer from its record of success with these that its implications for spherical diffraction will succeed. Here, the explanatory mechanism is the representation of the light source as a wave whose front is a source of secondary propagation, all of whose possible paths contribute to the intensity of the incident light. Treating this conception instrumentally provides no basis for predicting how it will fare in a novel application, but at least its reliability for the phenomena by which it has already been tested can be induced. Again, the instrumentalist may feel entitled to use special relativity to predict the outcome of a Kennedy-Thorndike-type experiment on the basis of the success of relativistic kinematics with Michelson-type interferometers, even if his refusal to treat realistically the relativistic conception of mass leaves him no grounds for predicting mass-energy conversion.[9]

This is a mistake. The problem for instrumentalism is worse than this line of reasoning imagines. For instrumentalism is subject to the argument in chapter 3 against the autonomy of enumerative induction as a warrant for prediction. I now, as promised, return to that argument as a stage in the argument for MER.

The difficulty lies in the logic of the "straight rule" of induction. Straight induction takes apparent regularities to be genuine. It projects a concomitance

---

9. The Kennedy-Thorndike interferometer experiment of 1932 has been invoked as an example of a test that is different in kind from those that originally confirmed relativistic kinematics, because in it a null result is not predicted. But from the standpoint of the criterion of novelty, the difference is incidental, as the Kennedy-Thorndike result instantiates the empirical generalizations on which Einstein's reasoning is based. Those who offer the Kennedy-Thorndike experimental design as a case of independent testability admit that they have no general criterion for the independence of a test. See Grünbaum, "The Bearing of Philosophy on the History of Science"; J. Leplin, "Contextual Falsification and Scientific Methodology," *Philosophy of Science*, 39 (1972):476–491; Ronald Laymon, "Independent Testability: The Michelson-Morley and Kennedy-Thorndike Experiments," *Philosophy of Science*, 47 (1980):1–38.

of properties in observed cases onto new cases. Its basic form is an inference from so many things with property A also having property B, to additional A's also having B. The strength of the inference increases with the exceptionless continuation of the concomitance. The induction may be either a singular inference to the preservation of the concomitance in the next case, or a general inference to its universality. Thus, either from $Ax_1 \& Bx_1 \&, \ldots, \& Ax_n \& Bx_n, \& Ax_{n+1}, Bx_{n+1}$ is inferred, or from $Ax_1 \& Bx_1 \&, \ldots, \& Ax_n \& Bx_n, (i)(Ax_i \rightarrow Bx_i)$ is inferred.

There are general reasons, originating in Hume's skeptical analysis of causation, to be wary of the legitimacy of the straight rule of induction. What, following Hume, is known as the "problem of induction" is basically an argument that inductive inference depends on causal assumptions that could only be justified inductively. We are inclined to reason inductively, Hume noted, only in the context of background beliefs that link the premises and conclusion of our inference in a causal way—minimally, by appeal to a general presupposition of the regularity of natural phenomena. But, Hume continued, neither specific causal judgments nor general principles of uniformity have any a priori guarantee; they are reached only by inference from experiences that instantiate them, and such inference must be inductive. Thus, inductive reasoning is circular. The legacy of Hume's argument is a skeptical challenge to show why the fact that certain things have been found to exhibit a certain regularity is any reason to expect that *different* things, about which there is as yet no information, will also exhibit this regularity.

The challenge can be taken up in a number of ways, and several "solutions" to Hume's problem have become standard.[10] None of them work; their interest

---

10. The solution that first comes to mind is an appeal to the success of science. The problem of induction is remarkable for its generality. It is insensitive to vast differences in modes of induction, undercutting, equally, the scientist's marshaling of evidence and the prognostications of the soothsayer. Surely, it is felt, evidence and superstition are not on the same epistemic footing, and any argument that applies indiscriminately against both must be fallacious in some way. Unfortunately, it is only by induction that the epistemic superiority of the imagined deployment of the straight rule by scientists is established. The comparison of modes must proceed by some standard, and it is unsurprising that if induction is the standard, then induction wins. The success of science cannot warrant induction, because the warrant for basing prediction on what has succeeded *so far* is precisely what is at issue.

More revealing is to use the problem's apparent intractability as an indication that it is misconceived, and to attempt a *dissolution* rather than a solution. If induction is an a priori standard of warrant, then it is unsurprising that no grounds for induction are discoverable. If arriving at one's expectations inductively is what it *means* to be justified, then a justification of induction is neither possible nor necessary. The question of why *induced* expectations are the right ones to have becomes analogous to the question why *moral* actions are the right ones to perform.

This approach ignores the difference between justification in the sense of conformity to abstract principles of reasoning, and prudential justification. It can be against one's interests, and therefore irrational in a prudential sense, to act rationally by general norms. It has been suggested, for example, that President Eisenhower's broken syntax and fumbling inconsistencies at press conferences were deliberate attempts to appear homely. (Would that such craftiness could excuse some of his successors.) But even ignoring this complication in the notion of justification, the proposal does not work. For there are different ways of applying the straight rule, yielding different outcomes, and the claim

lies in revealing the various forms in which the problem reappears in response to them. But one response to Hume that continues to claim legitimacy is to attempt to found a rationale for inductive inference on *deductive* principles, via the axioms of probability theory. This approach lives because it benefits the Bayesian. Indeed, it is supposed to be a unique advantage of Bayesian methodology to underwrite straight induction, by showing that the probability that the next A will be a B approaches unity, as does the probability of the universal generalization that all A's are B's, as the number of A's that are B's approaches infinity. It is argued that one cannot consistently maintain that the probability that the next A will fail to be a B exceeds any preassigned, positive number, however small—and thus remains above zero—as the number of A's conceded to be B's increases, without violating the axioms of probability.

Unmentioned in this argument is the fact that at no point in accumulating A's that are B's can one possibly have acquired a body of evidence on the basis of which the probability that the next A will be a B need be assigned a positive value, however small. Any finite sequence—and finite bodies of evidence are the best we can do—could be aberrant, in that the pattern it exhibits could radically misrepresent that to which the sequence converges in an infinite limit.[11]

Moreover, the deduction, from the axioms of probability, of conclusions about the next case, or the universal generalization, depends on assuming that the generalization is not assigned zero probability independently of empirical evidence. In Bayesian terms, what is supposed to happen is that as the evidence—in the form of A's being B's—mounts, so does the probability that the next case conforms to the evidence. We begin with some initial probability assignment, presumably low, to the universal generalization, and adjust this probability upward in response to new evidence. The Bayesian then proves that the probability of the universal can be brought arbitrarily close to unity by accumulating enough evidence. That is, no matter how close to unity one wants the probability of the universal to be, there is *some* amount of evidence sufficient to guarantee that the probability gets this close. That is,

$$\lim_{n \to \infty} \Pr((i)(Ax_i \to Bx_i)/Ax_1 \& Bx_1 \&,\dots,\& Ax_n \& Bx_n) = 1.$$

----

that induction is intrinsically rational underdetermines the choice among them. Hume's original question was what, from a given body of evidence, it is warranted to infer. The answer "infer what can be induced" is inadequate, because incompatible things can be induced. This problem is developed below.

11. John Earman and Wesley Salmon write, "Humean skeptics who regiment their beliefs according to the axioms of probability cannot remain skeptical about the next instance or the universal generalization in the face of ever-increasing positive instances" (Merrilee Salmon et. al., *Introduction to the Philosophy of Science* [Englewood Cliffs, N. J.: Prentice Hall, 1992], chap. 2). The trouble is that no person, Bayesian or not, can ever *be* in the face of "ever-increasing" positive instances in the sense that this assertion requires. No amount of evidence can establish that the positive instances are "ever-increasing."

The proof requires, however, that the initial probability of the universal—its Bayesian prior probability—be nonzero; otherwise adding evidence will not increase it. But the probability of the universal is the infinite product of the probabilities of individual cases—the probabilities that particular x_i's have B if they have A, given that x_j has B if it has A, for $j<i$; it is the limit of the product of these individual probabilities as their number approaches infinity. This limit is nonzero only under exceptional circumstances, circumstances which require the individual probabilities to be interdependent.

To see this, write the universal, $(i)(Ax_i \rightarrow Bx_i)$, as an infinite conjuction: $Ax_1 \rightarrow Bx_1 \& \ldots \& Ax_n \rightarrow Bx_n \& \ldots$. The probability of this conjuction is the product of an infinite sequence of probabilities. The product of an infinite sequence of numbers between zero and one will be zero, unless the elements of the sequence approach one; that is, the limit of the nth element of the sequence as n approaches infinity must be one. Nor does this condition ensure a nonzero limit for the infinite product. The sequence whose nth element is $n \div (n+1)$ approaches one fairly rapidly, yet the limit of its product is zero. We have seen that mounting evidence of A's being B's cannot impel a nonzero probability assignment to the next case. We now see that even ever-mounting probability assignments do not, in general, confer nonzero probability on the universal. To obtain a sequence whose infinite product is nonzero requires a very rapid approach to one. For example, the product of the first n elements of the sequence $\frac{3}{4}, \ldots, (2^n+1) \div (2^n+2), \ldots$ is $(2^n+1) \div (2^{n+1})$, which approaches $\frac{1}{2}$. Unless the terms of a sequence are fixed by preceding terms in such a way as to achieve a sufficiently rapid approach to one, the infinite product of the sequence will be zero. Needless to say, the terms of such a sequence are not probabilistically independent; their values are determined by their position in the sequence.

Only by assuming probabilistic interdependence can the probability of a universal be made nonzero, and even then it is nonzero only in exceptional cases. But to assume probabilistic interdependence is manifestly to beg the question against the inductive skeptic. Anyone dubious of straight induction, anyone unconvinced that the mere fact that certain A's have been B's is reason enough to expect other A's also to be B's, is hardly prepared to grant that one A being a B affects the likelihood that some other A will be a B. That the concomitance of A and B in one instance has any bearing on its likelihood in a different instance is precisely what is at issue in judging the legitimacy of reasoning by straight induction. The Bayesian's attempt to use the axioms of probability to underwrite straight induction therefore collapses. Its collapse was preordained. The Bayesian is in no position to deny the inductive skeptic's prerogative to assign a zero prior probability to the universal, since Bayesianism relies on a personalist interpretation of probability to supply priors.

The moral is that deduction from any evidence we could ever have cannot, in principle, underwrite induction. The conceptual difficulties afflicting straight induction must be addressed directly if the general reliability of a theory, with respect to a given domain of phenomena, is to be inferred from its record of success in that domain; they cannot be finessed by appealing to Bayesian principles.

Those difficulties are legion, and Hume's argument is at the heart of them. Hume discovered the unavoidability of empirical presuppositions in induction. More generally, if a *warranted* method of inference is one with a propensity to yield true conclusions over false ones, which would seem a minimal condition of warrant, then it is impossible to warrant *any* method of inference independently of empirical assumptions. *Deduction* has no greater propensity to yield truth than its premises have to be true. The one qualification to Hume's insight is that the form of inference to be warranted must yield conclusions to begin with. In a random universe, or one so perceived, induction has no propensity to yield any conclusions at all.

I shall discuss two conceptual difficulties of straight induction that display the perniciousness of Hume's discovery. The point in each case will be that the only way out is to rely on *other* modes of ampliative inference, which are not available to the instrumentalist. The conclusion will be that *we cannot be instrumentalists* if we hope to warrant the reliability of theories even with respect to the domains in which they have been tested and found successful. That will require realism.

One difficulty is the "paradox of the ravens." We observe lots of black ravens, and no others. Straight induction yields the universal "all ravens are black." But this is equivalent, by contraposition, to "all things nonblack are nonravens," which straight induction yields from observations of green emeralds (green emeralds being nonblack nonravens). But surely black ravens and green emeralds do not support inference to the same (to within logical equivalence) universal. Responses to this dilemma have focused on the equivalence condition—that logically equivalent conclusions be equally supported by the same evidence—maintaining that logical form affects inducibility. But, the desperateness of suspending the equivalence condition aside, this response misses the point. For green emeralds do not support the universal "all nonblack things are nonravens," let alone "all ravens are black." There is not the slightest plausibility in inferring that all nonblack things are nonravens from observations of nonblack nonravens. Why, then, is it plausible to infer from observations of black ravens that all ravens are black, or from observations of green emeralds that all emeralds are green? What is the difference?

The difference must be that color properties and natural-kind classifications are independently believed to be connected. "Nonraven" is not a natural kind, and "nonblack" is not a color. Only where a connection is assumed is there any inclination to infer by straight induction. Hume, motivated by specific examples, supposed that the needed connection is causal. A wider familiarity with natural law suggests that we identify it more broadly as *explanatory*. It is only where possession of property A helps us to understand why something should be a B that we are inclined to use the straight rule. In the event that A's persist in being B's without any such connection suggesting itself, we remain skeptical, demand further evidence, and inquire after possible indirect or recondite mechanisms, capable of linking A with B, that have not occurred to us.

Sophisticated statistical studies in the social and biomedical sciences frequently elicit these reactions.[12] Correlations observed in samples in highly controlled experiments are typically distrusted until someone adumbrates at least the rudiments of a process capable of explaining them. They are not generalized to the populations sampled until that has been done, and until the proposed process has been examined for consistency with established theory.[13]

This solution to the raven paradox protects the equivalence condition. Black ravens do support the nonravenness of nonblack things, but nonblack nonravens do not. We do not generalize on the nonblack nonravenness of green emeralds, because nonblackness and nonravenness have no explanatory connection.

Green emeralds are the subject of the second difficulty with straight induction, Nelson Goodman's "new riddle of induction."[14] Goodman despaired of finding a principled way to circumscribe the range of properties that may be reliably projected—applied to new instances of a kind to past instances of which they have been found to apply. Green is projectable for the kind emeralds, but not "grue," which is defined to be true of all emeralds examined so far and found to be green, and also of all emeralds not yet examined if they are blue. To project grue

---

12. I quote with approval from Stephen Weinberg, *Dreams of a Final Theory* (New York: Vintage Books, 1992), p.63: "Medical research deals with problems that are so urgent and difficult that proposals of new cures often must be based on medical statistics without understanding how the cure works, but even if a new cure were suggested by experience with many patients, it would probably be met with skepticism if one could not see how it could possibly be explained." Weinberg thinks it appropriate to dismiss as "mere coincidence" a statistical correlation, however strong, for which no plausible explanation in terms of deeper physical principles can be envisaged.

13. A number of studies have shown a correlation between prostate cancer and vasectomies in men, as reported in C. Mettlin, et al., "Vasectomy and Prostate Cancer Risk," *American Journal of Epidemiology*, 7 (July 1, 1991):107–109. The studies were not undertaken because of any specific suspicion of a relation between the two; on the contrary, there is no clear idea of how they could be linked. The studies were motivated by the importance of *any* information relating vasectomies and health *in general*. Researchers were not looking for increased rates of cancer, in particular, but were tracking the health of vasectomized men. When the correlation was observed, the response was not to conclude that having a vasectomy increases the risk of cancer, nor even to project the correlation onto the population at large, as straight induction would have it, but to stress the tentative and problematic character of the data, *because no theory was available to explain them*. The prevailing inclination is to attribute the correlation to incidental features of the samples—properties that do bear an explanatory connection to cancer and were not successfully controlled for. At the same time, some researchers set to speculating as to a possible theoretical connection. Should a plausible theory be proposed, the *existing* data will acquire inductive weight. Additional data are not required to effect this difference, though they will of course be obtained.

In other cases, no more and no better controlled data than has already been obtained for vasectomies and prostate cancer are enough for warnings to be issued and preventive measures taken, *because a theoretical connection has been made*. Thus passive exposure to smoking is held (by the surgeon general of the United States) to be carcinogenic, though the correlations obtained in almost all studies are below the norm for statistical significance; see "Respiratory Health Effects of Passive Smoking: Lung Cancer and other Disorders," Washington D.C.: U.S. Environmental Protection Agency notice 600/6–90/006F). Researchers describe their studies as having "linked" passive smoking to cancer and heart disease; they are prepared to speak of "links." But they do not describe vasectomy results in such terms. Unexplained correlations are not "links."

14. Nelson Goodman, *Fact, Fiction, and Forecast* (Indianapolis: Bobbs-Merrill, 1965).

is to infer from the greenness of examined emeralds that unexamined emeralds are blue. For an examined green emerald is grue, and an unexamined grue emerald is blue. Data showing green emeralds are equally data showing grue emeralds, yet we are willing to infer from that selfsame data that further emeralds will be green and *not* grue.

It is not just that grue is unprojected; it is *counterprojected*.[15] To the extent that emeralds are found to be grue we are confident that further ones *will not be*. This is a consequence of projecting green. Clearly, if any properties at all are going to be projected, then other properties, defined in terms of the projected ones in the way that 'grue' is defined in terms of 'green' and 'blue', cannot be projected, but must be counterprojected. So if we project, we must also counterproject. If straight induction is ever warranted, its counter is equally warranted.

The implication is that straight induction in and of itself, that is, straight induction as a *form* of inference, is *never* warranted. That A's have been B's *cannot*, in itself, be reason to expect A's to be B's. If we are to infer that an A will be a B, our warrant must lie in the particular nature of the properties that 'A' and 'B' designate, and not in the fact that our inference fits the pattern of straight induction. That pattern is no better than its counter.

It seems to me that this argument is definitive. Hume was right to deny that the fact that so many A's have been B's is any reason whatever to project B-ness among A's. The reason that we do not appreciate this is that no one ever reasons this way to begin with. We project not A or B, not variables, but specific properties selectively; adherence to a general rule of straight induction does not describe our ampliative thinking. The question is, What is the basis of our selectivity?

Goodman's suggestion was *entrenchment*; we project those properties that have acquired an established role in natural language. If it is ultimately arbitrary which properties have acquired such a role, language having conceivably developed very differently, then this suggestion abandons any claim of our inferential practice to being warranted. Unlike Goodman, I do not believe that natural language is so arbitrary as, but for fortune, to have entrenched grue rather than green. But I would not base selectivity on entrenchment in any case, since it is not sufficient for projectability. If entrenchment is simply a history of *use*, it is insufficient for projectability, because descriptive contexts can diverge widely. There are any number of cases in which regularities are not generalized, although their terms are entrenched in use and independently projected in other contexts.[16] If, on the

---

15. This is the important feature of "grue," and it is often missed. It is not just that we don't project grue, but that we can't, compatibly with what we do project. This suggests that we generalize the strategy. What we want is a predicate so defined as to defeat its own projectability—that is, one that cannot be projected at all. So we might use green*, stipulated to be a property that everything found to be green has, but that nothing else has. Then everything known to be green is green*, but nothing green but not known to be green is green*. So green* cannot be projected.

16. Presidents elected at twenty-year intervals, beginning in 1840, died in office, but no one would have thought Reagan less likely to die in office had the 1980 election been put off until 1981. A string of Democratic presidential victories accompanied Super Bowl victories by the NFC without affecting subsequent election odds. Rainfall in Michigan is correlated with the divorce rate in New

other hand, entrenchment is a history not just of use but of projection, then it cannot be the basis of projection initially. Entrenchment is not necessary for projection, either. An innocuous neologism like 'bred', defined as "blue or red," is projectable wherever 'blue' or 'red' individually is.

I base the selection of projectable properties on explanatory connections. The first step is to recognize that straight induction works only to the extent that other ampliative principles are deployed along with it. Without additional machinery, ampliative inference really would be the self-delusional hypocrisy that Hume attacked. The traditional addition is eliminative induction. A generalization is supported not directly by its own instances, but indirectly by their ability to eliminate rival generalizations and so narrow the field. The narrower the field of survivors, the more credible is each of them. As an observed green emerald is, equally, a grue emerald, it does nothing to favor the generalization projected from either property over that projected from the other. As between these options, it has no eliminative effect, and therefore carries no evidential weight on eliminativist principles.

Unfortunately, this process will never narrow the field enough to generalize, unless some further principle is invoked to restrict the initial range of alternatives. No amount of evidence will disqualify all gruelike variations on the properties we wish to project. Moreover, eliminativism as a strategy for warranting generalization must presuppose that some generalization among a range of prospective rivals is projectable. What is the warrant for this? We can imagine the field narrowing to zero if exceptions arise in tandem with proposals; what seems to be impossible is to narrow it to just one, as each survivor generates another under gruelike modification.

Eliminativism is innocuous enough, but it cannot take us far because the warrant it provides is purely negative. We need in addition, or instead, to require an explanatory connection between properties to be generalized. Only thereby do we get positive warrant for a generalization that survives elimination. To select *A* and *B* as projectable is as much to infer an explanatory connection between them as it is to infer their continuing concomitance. There can be no warrant to inferring one without the other. For if we did not believe that being *A* has anything to do with what makes something *B*, then we would not project *B*-ness among *A*'s even if it happens to have appeared among examined *A*'s. And we would not think that being *A* does have anything to do with being *B* if the properties did not cooccur.

Thus, the solution to the raven paradox is equally needed to make sense of the fact that of two incompatible conclusions, equally inferable by straight induction from the same evidence, one, rather than neither, is warranted. Mineral structure bears an explanatory connection to preferential reflection of light waves; it has none to time of examination. No theory connects the assortment of members of a natural kind by color with their assortment by moment of disinterment. Any

---

York State. These are *curiosities*. The relevant predicates defy induction not because they are cooked up like "grue," but because they are not plausibly connectable.

theory that did so would have zero probability on our background knowledge, and so be incapable of supplying warrant to our inferences.[17] We cannot infer such a theory, given our existing warranted beliefs, and so cannot infer the concomitance of such attributes.[18] Explanatory connections cannot be forged gratuitously. They require some basis in previously warranted beliefs. No doubt, if we were allowed to assume just anything, without the epistemic constraints of background information, we could construct an explanatory connection between mineral structure and time of examination. But if we could assume just anything, then neither straight induction nor any other form of inference would be warrantable.

The inference to a hypothesis from its explanatory resources is, I have noted, abduction. In reasoning abductively, we count the ability of a hypothesis to explain other things that we know as grounds for its support. Of course, there must be constraints on how much, and under what circumstances, explanatory resources support inference. But unless explanatory resources do count to some extent under some circumstances, positive instances do not support inference, either. For the difficulties afflicting straight induction can be overcome only by invoking explanatory connections. We cannot generalize a regularity that we cannot explain. Explanatory connections underwrite our use of the straight rule, so that we must be entitled to make them if we are to generalize. Therefore, we are warranted in inducing (by the straight rule, with or without eliminative help) only if we are warranted in abducing.

This reasoning assimilates the explanation of a regularity to the connection of the correlates within it. Is there no other way to explain a regularity? Perhaps the generalization of the regularity is itself an explanation of it, rather than an inference from such an explanation. Suppose that in sampling a population with a view to assessing its composition, I find that all the items obtained for my sample share a property in common—all the ravens I have examined are black, say. Does the hypothesis that all ravens are black not explain this? After all, if all ravens are black, then it cannot be surprising that I have found none of another color. The reason that my sample contains exclusively black ones is that the population from which the sample is drawn contains no other kind. Can I not then infer from the sample that all ravens are black, without having established, nor even imagined, an explanatory connection between being a raven and being black?

---

17. On Bayesian principles, such a theory would itself be incapable of acquiring warrant by instantiation. The Bayesian might appeal to this fact to account for the unprojectability of properties that lack an explanatory connection. But Bayesianism can give no reason why the lack of an explanatory connection, in itself, should matter. Bayesianism must either assign zero probability to all theories without foundation in antecedent beliefs, pronouncing them alike unsupportable by further evidence, or must extend personalist randomness to such theories. In my view, radically new theories, even theories that conflict with background knowledge, can be supported, despite having zero probability initially, if they forge a relevant explanatory connection. Bayesianism lacks the resources to discriminate between Goodmanesque combinations of attributes that cannot be projected, and merely unprecedented combinations that might be.

18. Of course, this situation could change. We can imagine it changing, or, rather, imagine *that* it changes. But only if it *does* change is there any consequence to warranted inference.

Earlier, I argued that generalizations do not explain their instances.[19] The present proposal apparently disagrees, and I am accordingly suspicious of it. I think that the difficulty resides in an ambiguity as to just what a generalization is supposed to explain. The hypothesis that all ravens are black may explain why black is the only color showing up in my sample, but it does not explain the regularity that the sample exhibits. Only described *as a member of my sample* is the blackness of any raven explained. The blackness *of these ravens*, which happen, incidentally, to have entered the sample (which is supposed, after all, to be random), is unexplained. The generalization of a regularity, *once revealed*, does not explain the regularity independently of its mode of revelation, because precisely the same explanatory question arises with respect to the generalization as arose for the regularity generalized. I want to know why blackness is accompanying ravenness; it does not help to tell me that it does so always. That it does in all cases does not explain why it does in some cases, but only extends this explanatory task. Moreover, the inference from my sample to the generalization is certainly abductive. I have not been shown a way to induce without abducing, and the remaining issue is only what conclusions are well and truly explanatory.

The conclusion that induction is warranted only if abduction is warranted may be highlighted by generalizing the uniqueness condition for novelty. Rather than stipulate that no other *theory* than that for which a result is novel provide a basis for predicting the result, it could be stipulated that there be no other basis *at all* for predicting the result. For if no other theory provides such a basis, then nothing does. In particular, the mere fact that the result has been obtained consistently under certain empirical conditions is not, in itself, an adequate basis for projecting it.[20]

A consistent skepticism that disavows all ampliative inference may be an option. A selective skepticism that proclaims the reliability of our theories, but disallows any possible grounds for accepting them, is not. Reliability can be projected only to the extent that it can be explained. Instrumentalism, *by itself*, contains no explanatory resources. Positioning the theoretical mechanisms used to predict observations beyond all possible epistemic access, it offers no understanding of why they should happen to work. Without such understanding, we have no grounds to project the continuation of their success. Instrumentalism gives us no more than a *record* of past successes.[21]

---

19. See chapter 1, under Excising Truth from Explanation."

20. I did not propose such a change in making this point originally (in chapter 3), because the argument for it had only been adumbrated. It is inadvisable even now, because the analysis of novelty should be able to stand independently of the realist agenda I mean it to serve. This agenda, in particular this agenda's opposition to instrumentalism, is the motivation for a stronger version of the uniqueness condition.

21. There is a textbook example of straight induction in which a computer, "Cassandra" by name, prints nothing but truths. After a long series of verified pronouncements, we come to trust her as an authority. We suppose that her further, as yet unverified, pronouncements are true, even though we do not know her basis for making them. The suggestion is that her record to date *justifies* this attitude. I suggest that if this scenario seems reasonable, it is only because Cassandra is described *as a computer*. That is, we attribute to her some source of information and a reliable method, only we are ignorant

In Defense of Abduction

As important as the interdependence of modes of ampliative inference is their distinctness. The point is to establish a mutual reliance between generalization and explanation, not to regiment their cooccurrence or collapse them into one another. Straight induction is not an act of abduction, nor need it accompany such an act. I require only the weaker thesis that straight induction would not, in general, be warranted, if abductive inference, in general, were not also warranted. What must be inferred along with an induced generalization is not a specific hypothesis explaining the concomitance of properties projected, but an explanatory connection of some sort between them. Examples that do not submit to straight induction show that to be prepared to induce, we must be equally prepared to infer that there is an explanatory connection; they do not show that we need be prepared to commit ourselves as to what precisely that connection is. We may be unsure of exactly what the connection is while being confident that there is one. The latter confidence is enough to support generalization. It is just that the observed concomitance of properties alone is insufficient to establish that confidence. We must also be able to adumbrate an explanatory link, to make plausible the supposition that such a link is present, by suggesting or conceiving a mechanism to forge it.

Despite its looseness, this requirement can be substantial, because in meeting it we are constrained by background knowledge, by the success of our theories. This constraint becomes ever more severe as science develops. Ancient peoples could induce with relative abandon, constrained by little more than their imaginations; we know better. But to satisfy the requirement is not to endorse a particular explanatory hypothesis. In the "grue" example, it is our confidence that *there is no such connection* that defeats the induction. If that confidence were undermined by some esoteric theory that claimed some basis in the rest of science, our inferential inclinations would adjust accordingly. We would not have to believe that the theory gets the connection right, only that there is one.

But if we believe that there is an explanatory connection, we can foresee endorsing a hypothesis as to what the connection is, once alternatives for it are narrowed and the explanatory link is tightened. Abduction, if not immediate, is in view. Similarly, we might be confident to the point of abductive inference of a specific explanatory connection, while yet unprepared to generalize on specific properties. Many factors could impede the continuation of an observed concomitance consistently with an explanatory connection between them having been

---

of what they are because her programming is unrevealed. We have a general idea of how she does it, based on our familiarity with the capacities of computers to store and process information. The example would not work if Cassandra were a witch who consulted the entrails of animals or a palm reader, because our background beliefs provide no basis for explaining how animal entrails or palm lines could be a source of information. Far from coming to trust a palm reader issuing truths, I'd suspect that I was being set up.

established. Mindful of complexities that arise in practice, we could be more reluctant to make specific predictions as to what will be observed in new cases, than to endorse an explanatory theory. We would, however, in accepting the theory, foresee the possibility of controlling for incidental influences to the point of making predictions, if not under actual conditions at least under idealized ones. Only if we do not think that generalization of observables is possible at all is explanation impotent to propel inference.

Accordingly, induction is warranted only to the extent that the supposition of there being an explanatory connection is also warranted. But how is that supposition, in turn, to be warranted? If its warrant must be inductive, then induction turns out to be epistemically prior to abduction. The question parallels the familiar one of whether it is theory or experiment that comes first. Traditional empiricism gives the priority to experiment. More recent philosophy of science, by emphasizing the theory-ladenness of observation, has created the presumption that theory comes first.[22] This presumption has been taken as license to import into the analysis of scientific change sociological and psychological theories of human behavior that take no account of distinctively scientific — or, more generally, epistemic — grounds for belief. But neither historical studies of science nor philosophical theories of language and meaning show that observation is any more theory-laden than theory is observation-laden. The phenomena themselves suggest, rather, a mutual dependence. The evidential significance of observations, if not their perceptual content, depends on the application of background beliefs and expectations created by the success of theories. But theories acquire these formative influences only to the extent that they are successful, and success is measured primarily in terms of experimental controls. Neither experiment nor theory has a clear priority. Induction and abduction exhibit a similar interdependence, and my purposes require no more complexity than this.

It is sufficient for me that inductive inferences warranted *now* depend on pre-established standards of explanatory coherence, which in turn may depend on what had previously been induced, and so on. But I would argue further that explanatory connections that underwrite inductive inference need not depend on antecedent inductions. There is more to causal agency than the observed concomitances that Hume thought exhausted our perceptual access to agency. The idea of agency is supported by elementary impressions of agency, as when we experience effort or pressure. Such experience is the primitive basis of the idea of force, which is undefined, or only implicitly defined, in classical physics.

A chalk mark made on a blackboard is not simply the concomitance of the chalk's motion with the appearance of the mark. One is directly aware of *making* the mark, of *imposing* that chalk upon the board. The mere cooccurrence of the appearance of the mark and the moving of the chalk is experientially distinguishable from the making of the mark. For that matter, one is aware of making the

---

22. See my discussion of Kuhn's views on the theoretical presuppositions of language, in chapter 3, under "Assumptions and Terminology."

chalk move, as distinct from intending it to move combined with its moving.[23] Hume almost admitted as much. His own analysis of causation as a habit of mind or psychological inclination induced by experience of past associations implicitly invokes the sense of agency that he sought to define away. For what is this process of induction whereby expectation follows upon repetition, if not agency at a psychological level? It cannot very well be simply another concomitance.

Examples of elementary impressions of agent causation do not establish that we make causal judgments before ever drawing an inductive inference, because more is involved in judgment than perception. But the possibility of an experience of agency unmediated by generalization from instances supports, at least, an epistemic parity between inducing and explaining. Agency is the paradigm of explanation.

Agency experienced is not agency inferred. But the ability to explain is as fundamental as the ability to recognize and project regularities. These abilities become increasingly problematic and inferential as their subject matter becomes abstract and theory-dependent. I recognize no point at which either inferring explanations or inferring generalizations gains epistemic priority over the other. Abduction is as fundamental a mode of inference as any. A conclusion need not be true, even roughly true, simply in virtue of having been arrived at abductively. But *no* mode of inference carries such a guarantee. Every mode of inference is suspect to some extent, as every mode of inference sometimes yields false conclusions. If error is to be avoided, so is inference. We could elect just to register the successes of theories and infer nothing from these successes at all. For many theories, that is probably all that we *should* do. Many ways of accounting for past successes carry no implications for further success, and with these we should often be satisfied. I have argued that novel success cannot be accounted for in any such way, but rather than account for it, we might attribute it to chance. A theory of any depth is bound to have novel consequences. It may seem unlikely that these will universally fail, whether there is any truth to the theory or not. Individual novel successes, even dramatic ones, are small inducement to abduction.

But in the face of a sustained record of novel success,[24] uncompromised by failure, and the continuing absence, despite prolonged effort, of rival explanations of novel results, one becomes impatient with such reticence. It is an option, to be sure, but what recommends it over the hypothesis that our theory has identified, with some accuracy, the entities or processes responsible for what we experience? It is difficult to specify with complete generality when abduction is warranted, but surely abduction is warranted under *these* conditions. For, recall that reticence with regard to representational accuracy *alone* is not an option. The only viable reticence is thoroughgoing, extending to the prospects for applying

---

23. "The experience of authorship, of doing a deed, is one of the primary constituents of our self-awareness," writes Roberto Torretti, in *Creative Understanding* (Chicago: University of Chicago Press, 1990), p. 272.

24. Recall, from chapter 3, that the *success* of novel predictions is understood to include their quantitative accuracy, although that is not a condition for novelty itself.

the theory in new areas, extending even to the projection of further predictive success of the sorts already obtained. If no representational accuracy is acknowledged, there is no basis for such pragmatic confidence either. It is difficult to pin down historically the significance invested by scientists in distinctively novel achievements, but there can be no question that science does not and could not operate with so obdurate an aversion to inference as must accompany antirealism under conditions of prolonged and disparate novel success. At this juncture, antirealism parts company with epistemological naturalism, and reveals its true roots in a foundationalist fear of error. It is simply too conservative to hold much attraction.

The realist alternative is to maintain that to the extent that a theory accurately represents the mechanisms responsible for what we experience while yielding novel predictions at all, it is likely to issue in a preponderance of true over false novel predictions. Even small inaccuracies of theory could subvert predictive success by happening to affect just those areas on which observable phenomena depend. And large inaccuracies could fail to be reflected in the observable phenomena, making the theory a better predictor than it has an epistemic right to be. Under such scenarios, predictive success is a misleading basis for abduction to the representational accuracy of the theory. But though these possibilities cannot be ruled out, they are a slim basis for disavowing inference altogether. It is reasonable to expect in general, if not unerringly, that representational accuracy accompanies predictive accuracy as regards novel phenomena, and that expectation supports abduction where novel success is sustained across a broad spectrum of experimental applications.

Problematic though the inference from novel success to theoretical accuracy may be, there can be no doubt that the hypothesis of theoretical accuracy renders a distinguished record of novel success comprehensible. It would simply be gratuitous to allow both that a theory is substantially correct and that its empirical credentials with respect to phenomena novel for it, as well as observable phenomena generally, are uniformly impressive, while persisting in regarding the later attribute as mysterious. Nothing could count as resolving such a mystery.

To desist from all inference to a theory's accuracy or reliability simply because of the possibility that the theory is massively wrong, and only happens not to reveal its unrepresentativeness at the observational level, is to set an unreasonable and unrealizable standard of epistemic warrant. Held to consistently, such a standard would be intellectually paralyzing. Commonplace beliefs that reflect ordinary norms of reasonableness are easily theoretical enough to fail it. Virtually all conditions ordinarily taken to warrant belief leave room for error. Mere possibilities alone do not defeat warrant by ordinary norms of reasonableness.

I submit that it is clear that engaging the issue in these terms gives the edge to realism. No specific realist commitments are yet favored, but a willingness, in principle, to extend credulity to theory becomes tenable. That is, MER is favored. But of course, the issue will not be engaged in these terms. Instead, counterarguments will be deployed. This is entirely appropriate; confidence in MER at this juncture is indeed premature. But the point is that the burden has now shifted. It falls on those who would deny the very possibility of warrant for theoretical beliefs. An argument for MER is on offer, and further resistance to realism must, whatever else it does, provide some basis for faulting either the adequacy or the quality of this argument.

The best of arguments is inadequate if its conclusion is independently impeached. Accordingly, some counterarguments will raise obstacles to realism in general, and counsel resistance to all theoretical belief as a matter of principle. As these arguments may be brought equally against any defense of realism, they are well known and easily anticipated. I defend realism, and MER specifically when necessary, against them in the next chapter. Criticisms of the particular defense of realism that I have proposed are of more immediate concern. These I can only hope to foresee, and I turn to them mindful that others may be better qualified than I to discern my vulnerabilities.

Novel Prediction versus Retroactive Explanation

The argument for MER depends on the uniqueness of representational accuracy—truth, in short—of a theory as an explanation of its novel success. To achieve this uniqueness, the analysis of novelty was designed to avoid a general sort of difficulty that plagues other approaches. Criteria for novelty in competition with mine fail to ensure its objectivity. In one way or another, they allow subjective or incidental, historically contingent factors to affect novel status. And if it is accidental whether or not a result is novel for a theory, or if novelty varies with choices among alternative perspectives that cannot be made in a principled way, then novelty cannot bear the epistemic weight I give it.

Thus if being novel is simply a matter of being new or unknown to the theorist, and the times at which results are obtained or learned are arbitrary over significant intervals, then novel results need not have been and results not novel might have been. If being novel is a matter of use in constructing a theory, then to the extent that there is leeway as to what results are used, novel status becomes arbitrary. A result can be novel even if a straightforward generalization of it, which some other result in fact prompted, was used. A criterion based on need, which seems more to the point than use, is equally infelicitous. If many results can be used to the same effect, none are individually needed and the floodgates on novelty are open. Under such conditions, whether or not a result is novel cannot make the difference between warrant for interpreting a theory realistically and warrant for attributing pragmatic virtues, let alone the greater difference between either of these and mere compatibility with the theory. These are the problems that motivate the unprecedented complexity of my approach.

But now the obvious question to pose is whether my approach solves these problems, whether it secures the objectivity and insensitivity to context on which the epistemic significance of novelty depends. The main source of reservation I expect is that I do not fully extricate novel status from historical contingency. This of course is true; the question is what contingencies remain and how pernicious they are. I have argued not that historical contingency, as such, is pernicious, but that particular historical or psychological factors built into other analyses are. These are ones I have sought to avoid. I have not avoided, and would not avoid, the contingecy that an independently discredited theory, could, but for the research agendas of certain scientists, have remained a viable alternative basis for predicting a result that I qualify as novel for a rival theory. I regard this possibility as innocuous on the grounds that its effect is to withhold novel status from a result that could have deserved it. Realists may object to this, but antirealists, against whose reasoning mine must be measured, should not.

Many realists are going to find my argument too restrictive; they will wish to be more liberal than I am in imputing epistemic significance to corroborating results. They may, for example, require not that a realistically interpreted theory claim *its own* record of novel success, but only that it bear certain logical or explanatory relations to a theory that does. They may want the relation of evidential warrant to be transitive across entailment, so that they will be entitled to treat realistically a theory that entails a theory that has novel success. They may defend this liberalization by pointing out that scientific practice does not limit the source of empirical support to predictions obtained directly from the theory supported. And this is but a minor concession among those some realists will demand. Many will claim a much wider legitimacy for explanatory inference than I defend. I wish these realists *bon voyage*. Theirs is a journey through troubled waters that I would moderate realism to avoid. If they make it, fine.

Then there is the supposed historical contingency that theories capable of predicting a result novel for an extant theory happen to await development. Had they been developed already, the result would not be novel. This possibility is more threatening in one way: It cuts oppositely, disputing an existing entitlement to wax realistic. What I have said about this worry is that it should bear on the epistemic significance to be invested in novelty, rather than on novel status itself. My approach is to protect the warrant for realist claims from their ineluctable defeasibility. The possibility that a claim will be defeated is insufficient to undermine its warrant. That argument will be continued in chapter 6. For now, I note that the unavailability of a rival to a theory that claims novel success *may not be a historical contingency*. There may be no rival, not because of where theorists happen to have directed their efforts or how resourceful they have been, but because the world does not admit of any rival that can pass empirical scrutiny. Admittedly, our failure to have developed a successful rival does not establish this, but it can be evidence of it. The general question of the availability of rivals will be revisited in chapter 6.

I find the most serious challenge to my approach's claim to objectivity in the importance that I attach to a theory's particular provenance. Were a theory's provenance incidental to it, the same theory having, but for fortune, emerged from a very different line of reasoning, invoking very different empirical generalizations, then novelty on my analysis is equally incidental. Suppose that the very theory to achieve a record of novel success that invites realist explanation had come along after the fact, and been developed expressly to account for results that happen to be novel for it. Or suppose a theory so developed had originated differently, so that results it merely reproduces come out novel. I acknowledge that these scenarios cannot be ruled out as a matter of principle. But I do not need to rule them out as a matter of principle, any more than I need to prove that a massively unrepresentative theory cannot be a good novel predictor. It is enough that scientific experience discredit such scenarios, and I submit that it does.

Even less ambitious visions than realism of what beliefs science is capable of warranting demand that the epistemic relations between theory and evidence be stable across alternative courses of inquiry in which the same facts are known and the same theories are proposed. There is an element of chance in the timing and circumstances of the introduction of theories. Historians tend to regard theories as creations of their times, as the inevitable outcomes of a certain mesh of social forces and intellectual interests, but such determinism is hard to square with the pivotal role of key individuals in producing new ideas. The best example is general relativity, which was almost entirely the work of one person.[25] In general, it seems easy to imagine major theories having come sooner or later than they did, or not having been proposed at all, if different individuals had been on the scene. Some historians pose as a deep question why relativity and quantum theory originated in Germany. I favor the obvious answer that Planck and Einstein were German.[26] What damage does the historical latitude for theory-origination do to the stability of the relation of evidence and belief?

I am concerned to defend realism only under conditions of novel success. For this reason, the possibility that relativity, say, had come later is of little moment for my argument. If results novel for a theory had been discovered independently, and the theory emerged subsequently from reasoning that relied on those results, we would then be less inclined to interpret the theory realistically. Playing out this scenario fully, there is no realism to defend. My argument is not thereby discredited; it is simply inapplicable. If the worry is that theories we do have reason to interpret realistically might, but for fortune, have been entitled to no such credence, my response is that it is then all to the good that things happened as they did, rather than as they might have. There is nothing epistemically troubling about beliefs enjoying an element of luck. But for a witness, a murder might have gone unsolved; this possibility does not call our system of justice into question. It is frequently but good fortune that we are in a position to reach the conclusions that we do. What matters is whether the position we are in justifies the conclusion we reach, not how we happen to be in this position. Those who would oppose my argument can take no solace in the possibility that the evidence I require for realism might not be available. If it is unavailable, then an opportunity to be realists has been missed, but opportunities are missed all the time. What matters is not whether the evidence is available, but whether it would support realism if it were available.

Despite its innocuousness, this is a worst-case scenario. We are asked to imag-

---

25. Admittedly, there were other approaches to the field equations of general relativity—most notably the approach of David Hilbert, who sought simply to implement general covariance mathematically. But Hilbert's results are, conceptually, far short of Einstein's theory; they do not attach pivotal importance to the principle of equivalence.

26. To this must be added the fact that the resources available in Berlin were sufficient to induce Einstein, a most reluctant German, to relocate there.

ine that a theory so develops as to rely on results that in fact are novel for it. The reliance imagined here must conform to the independence condition for novelty; otherwise, this revisionist history makes no difference to my argument. Had the theory simply taken advantage of additional relevant information without its provenance logically depending on that information, the novelty of its success would remain unimpaired. It does not defeat novelty, for example, that the theorist have advance knowledge of the results in question, or self-consciously set out to explain them. Getting Mercury's precession to come out right was, from the beginning, as I have noted, a condition of adequacy for Einstein in developing general relativity. Once again, it is not this that defeats the novelty of the result, but the possibility of a qualitatively adequate Newtonian solution. In general, that a result is explained retroactively is compatible with its novelty. All we need is a minimally adequate reconstruction with respect to which the result is incidental to the theory's provenance. A psychologically formative role is not decisive. I submit that the more plausible historical scenarios preserve novelty as the order of theory and results is imagined to change.

The opposite hypothetical scenario, which makes results not novel turn out to be, may seem more of a challenge. Relativity could have come sooner. It is a remarkable fact that the development of modern physics can be recounted, to good approximation, without mention of the contributions of the most important theoretical physicist of the turn of the century. There is a natural progression from Maxwell to Einstein that skips Lorentz, and there is little evidence that Lorentz's theory of electrons made much difference to Einstein's reasoning, even though Lorentz's theorem of corresponding states achieved near mathematical equivalence to special relativity.[27] The possibility of bypassing Lorentz is evident in the reconstruction of chapter 4.

This scenario does not seem to make any difference to novelty; for reasons already explained, Michelson's result would not gain in novelty by postdating relativity. But we can try to imagine a situation in which such a change would matter. What if a theory that relies on a result, in the sense given by the independence condition, had come before the result was even known? If the result then qualified as novel, it would be a matter of luck that we lack support for a realist interpretation of a theory that in fact deserves no such credence. The implication seems to be that theories in which we do place credence need not deserve it. If whether or not a theory gets interpreted realistically is a matter of chance, then what confidence can we have in the warrant for theoretical belief?

I am predictably skeptical as to the plausibility of such a scenario. Theories are not individuated by their provenances, but it is highly unlikely for the same theory to arise from a very different provenance, and still less tenable to suppose it coming out of nowhere in some unreconstructable flash of inspiration.[28] Sophisticated theo-

---

27. For details, see Leplin, "The Concept of an *Ad Hoc* Hypothesis."

28. I would remind probabilistic epistemologists who inveigh against such attributions of likelihood in the absence of statistical comparisons that probabilistic concepts originate in elementary impressions of likelihood, not the other way around.

ries have highly intricate lines of development, and we know what a difference it can make to the outcome for one key piece of information to be lacking.

The possibility of developing special relativity without reliance on the light postulate that Einstein took from electrodynamics is a worthy example to challenge my confidence. Of course, the importance of *c* in the theory is not that it is the velocity of light; its importance is its invariance. A more abstract approach in which *c* is not connected to any particular physical phenomenon was possible.[29] Then symmetry principles about space and time become relatively more fundamental, and the constancy of the velocity of light relatively less fundamental, in the final theory. In essence, what happens is that the derivation of the Lorentz transformations is severed from their origination in electrodynamics.

It is unclear, however, what difference this alternative provenance makes to the epistemic standing that my analysis grants the theory. One still needs some physical phenomenon—if not light, something else—to fix the value of *c* in the final theory. Is the worry to be that the behavior of light might become a source of novel confirmation that the theory, as actually advanced, does not enjoy? Does the Michelson-Morley result become novel in the alternative scenario?

I am willing to accept this implication.[30] It shows that the actual course of history can deny novel status to results that could have claimed it, and doing that *weakens* (by my lights) the case for realism. But the challenge to me must be that the case I use history to make is by accident too strong. Results that *are* novel on my analysis—like mass-energy equivalence for special relativity—would have to turn out not to be under plausible, alternative provenances. But mass-energy equivalence certainly retains its novelty on the supposition that relativity arises without reliance on the light postulate.

This reply may not alleviate the general worry that the possibility of there being lots more novelty than in fact there is makes actual novelty too contingent to bear epistemic weight. What about cases in which the additional novelty created by a hypothetical provenance makes a net difference to a theory's epistemic standing? The most extreme example to motivate detaching a theory from its provenance is afforded by contemporary speculation about the unification of fundamental forces through abstract symmetries. Theorists are wont to imagine that armed only with the right symmetry one could reconstruct the forces of nature without the benefit of experiment. Maybe we could get Maxwell without Faraday this way. The idea is that local gauge symmetries, together with principles of simplicity or economy, will indicate that certain forces are necessary to offset gauge changes that the symmetries say should not matter. Thus the symmetry of voltage changes requires the electromagnetic force; isotropic spin symmetry re-

29. This was first shown by W. von Ignatowski in 1910. See John Stachel's discussion in "Scientific Theories as Historical Artifacts," in Leplin, *The Creation of Ideas in Physics*. Significantly, Stachel concludes that such a difference of approach would have produced a "different theory."

30. In historical context, it would have to be the behavior of light that was used to fix *c*. The possibility of using, say, neutrinos is recognizable only retrospectively. So it can still be argued that the behavior of light is not novel under the alternative scenario. I do not avail myself of this line here, because I am trying to accommodate the champions of historical contingency as far as possible.

quires the weak force; the hadron symmetry of quark confinement requires the strong force; symmetries of states of motion require the gravitational force; and, ultimately, more abstract symmetries that subsume these require the unification of forces.

It is compelling to imagine deducing from a theory that postulates the invariance of an abstract mathematical quantity, without any clear connection to observable physical processes, the experimental effects that led to the development of earlier theories. From the perspective of a theory that depends only on the assumption that the world is deeply symmetrical, and that is capable of delivering all the forces of nature, virtually everything that happens is novel. What if such a theory had come first, and the empirical information that was in fact used to build physics was not needed and, for good measure, was not yet available? Then there would not be all the more limited, alternative explanations we have developed for particular results to obviate all that novelty. This possibility suggests that it is only an accident, on my argument, that our credulity does not extend far beyond what our experience in fact warrants. Perhaps, then, such credulity, if any, as my argument does warrant is only accidentally warranted.

Speculative theoretical physics poses deep challenges to our standards of warrant, possibly requiring, or at least justifying, the evolution of those standards into something different from, and probably weaker than, the standard of novel success. These challenges are the subject of chapter 7. But in defense of that standard here, I note that this troublesome scenario glosses over some serious obstacles. Most obvious is the question of *which* symmetry, at a level of abstraction that takes us beyond the theories that have developed from empirical information, is the appropriate one to assume. How is that choice to be made independently of empirical information? Once we move beyond electroweak unification, which claims the novel success of having produced new forms of light, to the level of grand unified theory (GUT) combining the electroweak with the strong force, there are an indefinite number of distinct symmetries to pursue. There is an experimental basis for disqualifying certain candidates; the simplest SU(5) symmetry is disqualified by the failure to detect proton decay, for example. But it is far from clear that abstract principles of consistency and simplicity suffice, in the absence of experimental controls, to identify a particular theory uniquely. There is some hope that this will happen, but that is not enough to sustain the challenge that concerns me.

Perhaps a larger obstacle is the status of the appeal to simplicity to begin with. It does not suffice to rationalize all the assumptions made in generating global theories. In $n = 8$ supergravity, which, until the resurgence of string theory, appeared the best prospect for a quantum theory of gravity that would unify gravity with the remaining forces, particle spins are confined to the interval $[2, -2]$. This yields eight particle-conversion steps, which reduce to one in 11 dimensions. This looks very simple. But is this simplicity enough to identify $[2, -2]$ as the right interval, or 11 as the right number of dimensions, in the absence of empirical considerations? At one level less removed, is it simpler to elect a symmetry that puts magnetism on a par with electricity, and requires magnetic current and magnetic monopoles in the manner of GUTs, than to make magnetism derivative and reducible to electricity in the manner of Maxwell? How can it be settled abstractly which of these, or other, courses nature should elect to follow? Scientific experience indicates that nature answers to aesthetic demands, but *which* aesthetic demands are the appropriate ones to make of her?

Theories require a heavy input of experimental information for their development. To imagine their provenance indifferent to specific empirical laws, much less to empirical information altogether, is fanciful. The problems and explanatory tasks for which theories are designed so influence the reasoning leading to them that it is implausible to imagine those constraints withdrawn while holding the resulting theories constant. To suppose that, but for fortune, any given theory could have issued from a provenance so different as to subvert claims to novel support is but stipulation. Perhaps the supposition can be made good in isolated cases, but as a general moral it is precipitous.

Apart from this cautionary response, there is a way to answer the challenge. Suppressing my impatience with alternative scenarios for theory development, I will examine the consequences of my analysis of novelty for some of the more plausible variations. Suppose that the provenance of theory T includes three results, any two of which would have sufficed as an empirical basis for generating T. As none of the results are individually indispensable, are all of them absent from a minimally adequate reconstruction of T's provenance, and so novel for T? To generalize, suppose that n results are used where any $n-1$ would suffice, for large n. It appears that the more the information needed to construct T, the greater is the novel success that accrues to T. T appears to have $n!+1$ possible provenances, individuated by the particular combinations of results used, with respect to all but one of which a different one of the n results is novel. If novelty is insensitive to contingent variations of provenance, do all n results qualify as novel? This is surely a disastrous consequence.

In this scenario, no premise is common to all of T's provenances. But a single empirical generalization is common to their minimally adequate reconstructions: the disjunction of all n conjunctions of $n-1$ results. The differences of provenance disappear under reconstruction and as a result, have no bearing on T's ensuing novel success. All n results appear in the logically weakest empirical premise needed for the reconstruction, and so *none* are novel. The supposed difficulty is not recovered by supposing that some one result, O, is essential to all provenances but insufficient, requiring supplementation by any of the results O_1, . . . ,O_n to generate T. Each O_i is individually inessential, but their disjunction is essential to every provenance and so appears as a premise in a minimally adequate reconstruction. Consequently, none of the O_i are novel. Nor would O become novel if the conjunction of the O_i were sufficient without it, for then the disjunction of that conjunction and the n conjunctions that conjoin each O_i individually with O, viz.,

$$(O_1\& . . . \&O_n) \vee (O_1\&O) \vee . . . \vee(O_n\&O),$$

would be the logically weakest premise. Generalizing on such scenarios, a result is essential to T's actual provenance in the sense of my analysis, only if it is essential to any adequate alternative provenance. Thus, a change in provenance does not, in general, affect novelty.

Admittedly, this conclusion does not extend to the more fanciful, wholesale divergences of provenance lately contemplated. But my analysis holds something even for those persuaded by that line. First, I would suggest weakening the requirement that evidential relations between particular facts and particular theories be constant across all possible changes of provenance. What really matters to the defense of realism is not that the warrant for theoretical beliefs be invulnerable to historical contingencies, but that *the property of being warranted* be thus protected. It is not necessary that a belief carry the *same* warrant across hypothetical scenarios, only that it continue to *be* warranted. Next, I suggest the plausibility of supposing that a theory with substantial novel results to its credit is likely also to claim novel success under hypothetical alternative provenances. It is just the particular results that come out novel for it that will be different. Its property of being epistemically warranted is preserved. Thus, if the behavior of light is to become novel on a more abstract scenario for the genesis of relativity, then the prospective novelty of whatever other physical phenomena are used to fix the value of c on that scenario is debarred.[31] The supposition that changing provenances does not affect the net level of novel support is plausible, because in constructing a hypothetical provenance it is insufficient simply to add results novel relative to the actual provenance; we must also replace or delete results already present. Simply adding results does not disturb the theory's minimally adequate reconstruction. To affect the reconstruction, we must suppose that results that were relied on are unavailable or unexploited, and different ones—presumably the originally novel ones—must then be supplied to complete the argument. The excised results are then candidates for novelty under the resulting reconstruction. The upshot is that the epistemic import I claim for novelty survives even if, contrary to my view, independent lines of argumentation, based on empirical assumptions whose qualitative generalizations are nonoverlapping, yield the same theory with comparable cogency.

Partial Truth and Success

My argument defends the inference from a theory's novel success to its partial truth, interpreted as degree of representational accuracy. Minimally, I am committed to the claim that the greater the novel success uncompromised by empirical failure that a theory sustains, the less likely are the theory's predictive and explanatory mechanisms to be wholly unrepresentative of the physical entities

---

31. As previously noted, this example is historically strained in the interest of meeting my contextualist critic as much of the way as possible. It would go too far to imagine all the physical processes that could, with hindsight, be used to fix c disjoined in the reconstruction of the more abstract provenance, precluding the novelty of each. Special relativity had already to be developed before the investigations that revealed the alternatives were possible.

and processes actually responsible for the relevant observations. I am vague by default as to how much novel success merits what level of confidence in representational success. Some technical comparisons are obtainable from consistency considerations, but I shall not pursue them as they would assuage only the appearance of vagueness, not its substance. Genuinely quantitative measures are no more obtainable here than they are in Bayesian confirmation theory, which must replace its variables by numbers that are totally arbitrary. Still, I shall try to be a bit more detailed as to how I would have my minimal claim understood.

I am not suggesting that the amount of novel success provides a measure of the degree of representational accuracy achieved. To explain novel success it is not necessary to attribute a particularly high degree of accuracy, because we have no way to determine what forms of inaccuracy might be irrelevant to the observable situation. The world could be far richer than our access to it can measure. I see no way to attach probabilities to such possibilities. I claim only diminishing probability that the richness that transcends our evidence violates constraints imposed by a theory with mounting support. Such richness might fail to be represented at all; our only epistemic handle is on the prospect that it is *mis*represented. We can, in principle, be warranted in believing something theoretical about the world, even if we cannot, in principle, be warranted in believing how much that something is on any absolute scale. The concept of representational accuracy does not supply an absolute scale, because it does not depend on, or even allow, any coherent notion of complete accuracy. My argument supports no notion of any theory's being "wholly true," let alone of there potentially being any theory true "of the whole." In fact, I find the idea of assessing a single theory as "true of the whole" incoherent, simply on the grounds that no determination of what "the whole" includes could be made independently of the theory itself. Applied to physical theories, such expressions are extended beyond any context in which they have a clear sense.

The expressions "theory of everything" and "final theory" have acquired a sense in contemporary theoretical physics, but the sense they have acquired is relative to an implicit circumscription of the potential range of our empirical knowledge. Such theories do not pronounce as to the extent of that range in absolute terms. A system of equations with which everything that happens, observable or not, must comply need not be the whole story, however much empirical content they add to the logical laws that begin with that status. For such equations could themselves derive from something more fundamental. There are reasons to anticipate limits to the potential depth of theories. At the range of 10^{-29} cm, space-time is supposed to take on a foamlike character, and quantum indeterminacy precludes further levels of structure. It would seem that a theory with such consequences is capable of forecasting its own completeness. But our epistemic attitude toward such a theory must still be hedged to reflect the possibilities of incompleteness and systematic misrepresentativeness in our evidence. Even a system of equations, as yet elusive, that claimed uniqueness on the basis of consistency considerations alone could not logically forestall its own revision. For, as we have seen, consistency conditions capable of imparting uniqueness of theory choice in contemporary physics include symmetry assumptions that have no a priori warrant.

I am also not claiming that a partially true theory is likely to be successful. Partial truth could be more probable than utter falsity given success, without the probability of success given partial truth being high. For partial truth to explain success, and for this explanation to support an inference to partial truth, does not require that partial truth render success probable. A genetic defect can explain a disease and be inferable from it, even if most people with the defect do not contract the disease. What I do claim is that the probability of success given partial truth, however low, is higher than the probability of success given utter falsity, relevant auxiliary knowledge and technological resources being held constant.

Indeed, this is something that I must claim. Writing '*PT*' for partial truth, '*UF*' for utter falsity, and '*S*' for success, my argument maintains that $pr(PT/S) > pr(UF/S)$. By Bayes's theorem, this requires that $pr(S/PT) \times pr(PT) > pr(S/UF) \times pr(UF)$. It is plausible to assume $pr(UF) > pr(PT)$; the prior probability of *PT*, absent *S*, is presumably lower than that of *UF*, if only because ways of going wrong would seem more profuse than ways of being right. I must claim a corresponding excess for the likelihood of *S* with *PT* over *UF*.

Given a propositional view of the nature of theories, we can distinguish two dimensions along which a theory's partial truth may be estimated. I have appealed so far to the relative representational accuracy of the account of explanatory and predictive mechanisms that the theory's propositions give collectively. It is also possible to attribute truth simpliciter to some among the theory's propositions, and to make the proportion of these among the total system of propositions a measure of partial truth for the theory. This still does not provide an absolute measure, for two reasons. One is that theories are indefinitely extendable. They can be supplemented and refined to apply in new situations, or in situations that we learn to describe in less idealized, more realistic ways. The other reason is that a theory's propositions differ in degree of centrality and in the depth of detail of the information they provide. An absolute measure would require a circumscription of a theory's total content, and rankings as to centrality and depth for its constituent propositions. These requirements cannot be met in any general way. A circumscription of content would be arbitrary, or at least temporally indexed to the state of relevant auxiliary knowledge. And any meaningful rankings would be specific to the theory.

I can, however, suggest a basis for relative estimates along the second dimension of partial truth. I have raised the possibility of individuating theories by identifying certain of their propositions as basic, in the sense that any departure from them would count, in the judgment of the relevant community of practitioners, as theory abandonment rather than theory revision. I have mentioned such examples as Newton's laws of motion and gravitation, Maxwell's equations, Lorentz's hypothesis of electroatomism, and Einstein's principles of relativity and of the independence of the velocity of light. One can interpret a sustained record of novel success uncompromised by empirical failure not only as indicative of representational accuracy for a theory as a whole, but as warrant for the unqualified truth of certain propositions that are basic within it, or, at least, of propositions that are implied by propositions that are basic. In the example of Eddington's light-deflection test of general relativity, which was interpreted more readily as warrant for that theory's law of light deflection in a gravitational field than as warrant for the theory as a whole, the unqualified truth of the law was taken to be the object of warrant. Some of a theory's novel success may credit certain of its components specifically in an unqualified way. As more components acquire such warrant, the theory's truth status improves. It is further possible to verify, or sustain, judgments of partial truth made on this basis by monitoring the survival of warranted components when the theory as a whole is rejected and replaced by a new theory. This process will be important to the defense of realism in chapter 6.

This second dimension of partial truth seems more concrete and accessible than the first, but it is also more speculative and problematic. It is not clear, in general, how to distinguish warrant for a theory as a whole from warrant for limited components, nor warrant for truth, as such, from warrant for imputing some degree of accuracy. Nor is it clear how to distinguish in general terms between unqualified warrant for the partial truth of a theory or components of it, and partial warrant for the unqualified truth of components. Nor do we have a general criterion for classifying laws or hypotheses as components of a theory, as against empirical consequences of a theory. One may endorse a law of gravitation as a description of the motions of observable bodies, without endorsing any explanatory theory of gravitation. While warrant for specific hypotheses may be the plausible finding in certain examples, my argument does not show how in general to support any totally unqualified realist claims.

There is, however, one advantage that I do claim for the second dimension of partial truth: It provides a way of understanding *total* falsity, or complete representational inaccuracy. So long as we think of representational accuracy solely as a matter of degree, total falsity is as elusive as total truth. If it makes little sense to think of a representation as unimprovable, the opposite status is also slippery. It would seem that we can always get farther afield from what is being represented, as well as closer to it. Simply by giving further detail to a theory that is already on the wrong track, we increase its *misrepresentativeness*. For example, the positing of ever-more-detailed mechanical structure for the ether in mid-nineteenth-century electromagnetic theory would seem to have taken this theory further from the truth. It looks like the open-ended character of truth applies equally to falsity, and a finding of unqualified falsity would be as problematic as a finding of unqualified truth.

However, the rejection of theories on the basis of anomalies or counterevidence seems, in general, to be much more definitive than the support we acquire for them. Not only do we pronounce theories unsupported; we also pronounce them wrong. And sometimes, in doing this, we regard the theory as totally off base, not merely wrong in some respects or inadequate in some contexts. This is more likely to be the attitude in cases where competition among coeval theories is decided by strong negative evidence against one of the rivals, than in cases where a successful theory eventually gives way to a better theory through a smooth transition that weighs relative advantages and disadvantages. Thus, in phlogistic chemistry and ether-based electromagnetism, the central explanatory concept of the theory was definitively rejected in favor of a radically different approach. Phlogiston and the ether were not redescribed by subsequent theory, but eliminated entirely.[32] In such cases, we should regard additional refinements of its central concept not as increasing the theory's degree of falsity, but as adding unqualified falsity to it. It is helpful to have available a general conception of what we are claiming warrant for inferring when a theory is thus rejected. We may say that a theory is wholly false if its basic propositions contradict the basic propositions of a theory that is partially true.

My account of the justification for regarding a theory as partially true is, indirectly, an account of theory rejection. But the full story of rejection requires, in addition, an account of the force of negative evidence, an account that is sensitive to the difference between judgments of partial and unqualified falsity. I shall have more to say about this difference in chapter 6, in connection with the question of the epistemic status of rejected theories that achieved novel success.

The Pragmatics of Partial Truth

My notion of partial truth is metaphysical, unavoidably so as the notion of truth is unavoidably metaphysical. But when scientists hedge their bets by qualifying the conclusions to which the empirical success of their theories leads them, it is not a metaphysical point that they are making. Sometimes they do make metaphysical points, usually—when beguiled by philosophical challenge, public-relations pressure, or a socially induced propensity for modesty—instrumentalist ones that their working attitudes belie.[33] But something altogether different is going on when they appraise, not theories in general or the scope of scientific knowledge in the abstract, but specific theoretical conclusions. Their point has to do with the fallibility of the best warranted beliefs. Even if the notion of partial truth is itself metaphysical, the point of qualifying imputations of truth usually is not. Since realism in general depends on some notion of qualified truth, whereas

---

32. On criteria for distinguishing theory changes that preserve reference through redescription from changes that abandon reference, see J. Leplin, "Is Essentialism Unscientific?", *Philosophy of Science*, 55 (1988):493–511.

33. It is amusing to juxtapose Stephen Hawking's official instrumentalism, frequently appended to his popular writings, with the confident realism to which he unabashedly commits himself throughout the substance of his texts. In *A Brief History of Time* (New York: Bantam Books, 1988), he says both that theories are only rules relating observable quantities and have nothing to do with any reality outside our minds (p. 9) and that we have "theoretical reasons for believing that we have knowledge of the ultimate building blocks of the universe" (p. 66). In *Black Holes and Baby Universes* (New York: Bantam Books, 1993), he answers criticism of his use of complex, or imaginary, time to avoid big bang cosmology, by claiming that imaginary time has nothing to do with reality (p. 46); then he argues that it is imaginary time, rather than real time, that is really true of our universe, and he compares resistance to imaginary time with resistance to believing that the earth is round (p. 82). Of course, it is unrealistic to expect physicists, engaged in philosophical reflection, to meet philosophical standards of consistency and precision. But these vacillations are really amazing, and Hawking is by no means atypical. What accounts for such confusion?

the analyses of approximate truth and verisimilitude that philosophers have used have a checkered history,[34] I wish to say something about how *imputations* of partial truth can be understood in the context of my argument.

The matter has some urgency, because of the stringency of the conditions that I require for warrant. I have given many examples of novelty, but have not argued that the overall record of any specific theory is so impressive as to sustain a realist interpretation. No such argument is necessary to establish MER. However, scientists do frequently accept theories, and if mine is the correct analysis of what warrants them in doing so we should expect to find many cases in which the stringency of my standard of sustained novel success has been met. The alternative is to regard much scientific practice as precipitously credulous.[35]

Surely, my standard has been met by basic theoretical beliefs that have been preserved through theory change to become the common presuppositions of theorizing in diverse domains of inquiry—beliefs as to the existence and basic properties of elementary constituents of matter, for example. But with respect to many episodes of theory evaluation, the answer I propose is that acceptance represents a very limited commitment. There is reason, in the form of novel success, to believe that a theory has achieved some measure, possibly very small, of truth, and, on this basis, to pursue it: to refine or extend it, to apply it to outstanding problems, to base further theorizing on it. Seldom does acceptance amount to much of an epistemic commitment. It is primarily a pragmatic commitment to act on the theory as if it were true. Such a pragmatic commitment requires an epistemic basis; this is the upshot of my defense of abductive inference. But, as that discussion shows, the epistemic commitment that grounds a pragmatic commitment may be rather weak.

I think that what theorists have in mind in so limiting or qualifying their epistemic commitments is the likelihood of eventual theory change. They know that very impressive theories have come to be rejected, and they elect, accordingly, a measure of commitment to theories that impress them now that is compatible with a future finding of considerable error. So, to impute partial truth is to acknowledge up front that however good a theory looks to us now, we cannot expect it to be the final word.

But this is only part of the story. The idea of partial truth really points in two directions, and its import differs between them. When imputed to current theories that there is some reason to believe and no reason, as yet, to disbelieve, the emphasis of "partial truth" is on truth, and the intended contrast is with falsity.

---

34. There are several extant approaches to analyzing approximate truth, none of which I find helpful. In general, they are either limited to the empirical consequences of theories, in the tradition of Karl Popper's notion of verisimilitude, or they presuppose a realist framework and measure degrees of truth against presumed unqualified truths that are not epistemically accessible.

35. Larry Laudan has challenged me on this point, suggesting that it leads to relativism. Although congenial to relativists, unjustified theoretical beliefs among scientists do not seem, to me, to establish relativism. There may be absolute standards of warrant with which flawed individuals do not uniformly comply. But the question is important in its own right. If scientific practice often appears irrational, naturalism gives us cause to question whether we have identified the correct norms of rationality.

The claim is that in view of its success, the theory is more likely to be true, albeit only partially, than false. The claim is that the theory is not on the wrong track, is not basically misleading as to what entities or processes are actually producing our observations, and pursuing it will be progressive. The expectation is that even though the theory will probably be found to be wrong in important respects and will have to be replaced, it will not be judged, retrospectively, to have been misguided or counterproductive. With respect to a successful current theory, the "partial" in imputations of partial truth is a kind of generic reminder of the irremediable defeasibility of theoretical judgment; it does not respond to specific concerns about the particular theory.

In the retrospective orientation, applied to theories that were once successful but are, by now, supplanted, the emphasis of "partial truth" is on its being partial, and the intended contrast is with unqualified truth. The point of imputing partial truth in this direction is to emphasize that the theory's truth status was *only* partial. Here the focus is on specific ways in which the theory has been found to be faulty or misdirected.

Why not abandon truth claims altogether for rejected theories? Why think in terms of truth at all? We can do justice to a rejected theory's continuing pragmatic utility and predictive accuracy within certain ranges, without bringing truth into the picture. One answer is the disinclination to make an exception of current science. If we do not expect current theories to be the last word, then the attitude we take toward them should not be fundamentally different from the attitude we take toward theories that once looked as good as current theories do now. But this is a very general, philosophical consideration, a consistency constraint that does not speak to the details of what a retrospective imputation of partial truth intends to endorse. A more insightful answer is that the status of partial truth registers a certain view of the importance of the rejected theory as a contribution to the further progress that current theory represents.

Consider a homely example.[36] The power goes out in my house. I look out the window and see a utility truck parked nearby and men digging in the yard. I infer that telephone repairmen, responding to my earlier summons to correct a problem with my telephone, have inadvertently cut the power line to my house. In fact, it is not telephone repairmen who have cut the line at all, but cablevision repairmen whom I had not expected. My "theory," my explanation at least, of the power failure is false. Yet I am inclined to regard it as approximately true. It is closer to the mark than, say, the hypothesis that a power surge from an electrical storm is at fault. The importance of the difference is that I know where the problem lies and where to seek redress. I will just go out back and have a word with these gentlemen. The latter hypothesis would steer me wrong. It would be unavailing to summon an electrician to replace my old wiring.

In calling a past theory "partially truth," scientists are claiming that the areas

---

36. This example was suggested to me by comments my colleague John King delivered on a lecture by Arthur Fine ("The Natural Ontological Attitude") attacking realism's use of explanatory inference. Fine's argument will be considered in chapter 6.

in which it proved wrong are less important, in a certain respect, than the areas in which it was right. The extent of its representational accuracy can be measured only from the perspective of current theories, and that perspective is never unqualifiedly endorsable. Any measure of representational accuracy must presuppose current science, and so the representational accuracy of current science is not itself measurable. This implies that our retrospective measures can claim no absolute standing. Accordingly, partial truth, as an evaluation of past theories, has a different, nonmetaphysical import. It suggests that a theory's failings are not such as to compromise its ability to have advanced our theoretical interests, judged from a contemporary perspective. Partial truth, in this sense, is relative to interests, and imputations of it are revisable as interests change with new directions of theorizing.[37] Current science identifies certain interests: It invests special importance in the solution to certain problems, in the understanding of certain phenomena; it attaches special significance to certain observations. To judge a past theory to have been partially true is to judge that its failings are not pernicious with respect to those interests, that it turned out on balance to be a good thing, from our current vantage point, to have given that theory a positive epistemic evaluation and to have pursued it, despite its mistakes.

The difference between a judgment of partial truth and a judgment of outright falsity depends on whether or not the ways in which the theory departed from the truth, as currently reckoned, are important to our current interests. If believing the theory, rather than what we now believe, would not radically alter current directions of theoretical work, but only set that work back a ways—would not mislead us but only lead us less far—then the theory's falsity is less important to us than its truth, and it makes more sense to regard it as partially true than as simply false. Thus, we are not inclined to regard geocentric physics or phlogistic chemistry as partially true, but we are inclined so to regard Newtonian theory or special relativity. The advances to which the latter theories contributed outweigh their mistakes, from our current perspective.

This conception of partial truth gives a sense to truth comparisons and degrees of truth that we cannot get a handle on metaphysically. To the extent that the difference that the difference between believing the past theory and believing current theory makes to what we value now, to our pursuit of current interests, is slight, the past theory is truer. Degrees of truth for theories are inversely proportional to the degrees to which believing them would disrupt the pursuit of interests that current beliefs lead us to have. Judgments of degrees of truth on this conception are liable to change, but so are judgments of truth simpliciter. This conception enables us to make as much sense of judgments of partial truth as we can make of truth. It complements the conception of partial truth involved in interpreting theories realistically, by adding a respect in which judgments of partial truth can be comparative. Thus it does greater justice to actual scientific

---

37. Consider the change in the estimation of Cartesian physics occasioned by relativity theory. Because general relativity is reminiscent, in some ways, of Descartes's plenum theory of gravity, Descartes gets better press today than he did during the Newtonian hegemony.

reasoning than can the notion of truth as accuracy of representation alone. It pertains to the pragmatics of science, rather than to its epistemology.

I would emphasize the dependence of this way of understanding partial truth on contingent features of the state of science. For retrospective imputations of partial truth to make sense, current science must identify, clearly and uncontroversially, future directions for its own development. For what we judge progressive in the past depends on what we now take to be the right directions for further progress. To distinguish falsity from partial truth in the pragmatic sense, we must first identify the theoretical interests of current science. To the extent that these are in doubt or controversial, the evaluation of past theories is similarly afflicted. It is only because of its relative stability and its consensus over fundamentals that contemporary physics is able to recognize any of its history as progressive.

If physics really underwent the sort of revolutionary change that Kuhn has popularized—wholesale disruptions of both method and content that render the theories they divide incomparable—there could be no clear and stable judgments of the progressiveness of past physics. Not only would past theories be individually rejected; judgments of their relative worth would also be unstable.[38] Kuhn's own analogy to political revolution is informative here. Political revolutions characteristically rewrite history, turning heroes into villains and back again. Nothing like this is found in physics. As firm as the judgments that past theories were wrong are judgments that certain among them improved on their predecessors. Evaluations of past theories as progressive and partially true are as reliable as falsifications. This is evidence that theory change in physics is not revolutionary change of the sort that Kuhn depicted. By contrast, scientific fields that have not reached consensus over fundamentals, that have not established clear directions for further theorizing—fields in flux and controversy whose applications are disputed by experts—cannot coherently evaluate their own histories and are constantly switching heroes.[39]

---

38. Kuhn seems to recognize this implication of his position. He says that progress from Aristotle to Newton to Einstein is limited to scope of application and cannot be understood in conceptual or ontological terms (see *The Structure of Scientific Revolutions*, pp. 206–207). The point, however, is that physicists themselves certainly see the progressiveness of these transitions in conceptual terms, and should be unable to do so if Kuhn were right.

The stability of retrospective judgments of the progressiveness of theory transitions is an important counter to the skeptical historical induction on the failures of past science. See chapter 6.

39. My pragmatic view of partial truth bears some similarity to Richard Miller's account of approximate truth. The closest thing to my view that I have seen is Miller's remark, in his review of Arthur Fine's *The Shaky Game*, that we judge the importance of respects in which a theory is true partly on the basis of "the impact of the theory on subsequent improvements in science" (Miller, "In Search of Einstein's Legacy," *Philosophical Review*, 98 (1989):215–239).

6

Counterarguments

Overview

As I said in chapter 5, I expect resistance to MER to come not primarily from objections specific to my argument, but from independent objections to scientific realism in general. To the extent that these objections are influential, my argument is liable to be dismissed as a special case of an error already diagnosed, and, as such, as unworthy of detailed consideration. In this chapter, I want to respond, on behalf of MER, to some of these challenges. To maximize scope of application, I will structure my response, so far as possible, as a defense of realism generically, deploying the special resources of my own position where realism in general proves vulnerable.

For reasons that will emerge, I take most seriously the challenge represented by the "skeptical historical induction." A number of philosophers have argued that the history of science presents a record of theoretical failure that cannot be squared with implications of realism. The general form of epistemic realism, of which MER is a weak and limited instance, maintains that the empirical success of theories warrants crediting them with referential success and approximate truth. These attributions do not require that a theory survive all subsequent testing and theorizing to become a permanent part of the scientific corpus. But they do seem to require that a theory not turn out to be completely wrong. The realist should expect the central theoretical posits and hypotheses of an empirically successful theory to be retained in subsequent successful theories. From the perspective of current science, the central theoretical entities and their descriptions in past science, *insofar as past science was empirically successful,* should continue to be endorsed or, at least, to be recognizably recoverable. If they are not, then evidently the grounds for inferring them in the first place were inadequate. If such failure—such loss through theory change—is found to have occurred repeatedly, then empirical success is systematically unable to support a realist view of theories.

In his classic paper "A Confutation of Convergent Realism,"[1] Laudan adduces a series of historical theories, each of which he claims to have enjoyed, at one time, empirical success as compelling as any that the realist could want, but none of which contributes to the theoretical picture that science currently advances. Whatever grounds we now have for believing any theory can, Laudan suggests, be found, in equal or greater measure, to have supported theories that we no longer believe. He argues inductively that our current picture is unlikely, in retrospect, to have achieved referential success or approximate truth from the vantage point of future science, and that realism, according, has no application. The message is one of skepticism with regard to theoretical beliefs. The empirical achievements of theories exhibit a kind of cumulativity as science grows,[2] but the theoretical explanations of those achievements merit, at best, an instrumentalist or pragmatic reading.

A more widely influential challenge to realism is to be found in the thesis of empirical equivalence (EE), and the related doctrine of the underdetermination of theory by evidence (UD). EE states that every theory has rivals that do not differ from it—nor, therefore, from one another—at the observational level. With respect to whatever is, in principle, observable, the implications of the rivals are identical. Chapter 1 promised an examination of EE and UD. If EE is correct, then conformity to observation is impotent, by itself, to qualify any theory as deserving of realist interpretation. This would require some further principle of selection, beyond conformity to observation, to discriminate among the alternatives. UD then maintains that any such further principle would be nonevidential; perhaps such considerations as simplicity or utility can single out a particular theory as preferable, but there is no reason to hold that a theory thereby preferred is likely to be true. Simplicity, utility, unifying power, and heuristic fertility are examples of pragmatic virtues, in that they recommend theories for their relations to us rather than to the world. That a theory serves distinctively human needs is no indication that it accurately represents the world as it exists independently of human beings; consequently, pragmatic virtues are nonevidential. According to UD, no body of evidence is ever sufficient to warrant acceptance of any theory. Theory acceptance must be justified, in part, by interests that are irrelevant to truth and cannot, therefore, amount to a realist endorsement. EE and UD are building blocks of Quine's epistemology,[3] and they figure in the antirealist arguments of van Fraassen and Fine.

1. Published in Leplin, *Scientific Realism.*

2. Laudan develops his thesis of cumulativity at the empirical level in *Progress and Its Problems.*

3. Quine regards EE as a point of logic grounded in the Löwenheim-Skolem theorem. The Löwenheim-Skolem theorem guarantees the multiplicity of denumerable models for formal theories

A related challenge claims that explanatory power is nonevidential, not only because of its pragmatic character but also for methodological reasons.[4] Van Fraassen argues that the hypothesis that a theory gets all the relevant observable phenomena right is always preferable to a realist hypothesis that endorses the explanations the theory gives of those phenomena. He points out that the former hypothesis is less committal epistemically, and he claims that nothing more committal is required to make sense of scientific method. The ability to explain is pragmatically valuable. But because explanations are achieved at the expense of excess theoretical commitments beyond what the hypothesis of empirical adequacy—conformity to all observation—requires, they cannot be a source of epistemic warrant. The more and the better that a theory explains, the more valuable it is pragmatically but the *less* likely it is to be true. More generally, the "theoretical virtues"—what we value theories for beyond getting the observable phenomena right—vary inversely with credibility, because credibility varies inversely with content beyond the observational level at which content is epistemically unproblematic. Theoretical virtues cannot be truth indicators, because they increase epistemic risk.

Finally, an argument due to Arthur Fine creates an apparent dilemma for any realist view of science that relies on an abductive defense.[5] I shall formulate the dilemma a bit more broadly than Fine does, so as to connect it with the nonrealist view he favors. The dilemma begins by noting the prevalence *within science* of abductive methods. In science, we frequently find theoretical hypotheses being advanced or withdrawn for essentially explanatory reasons. Steady-state cosmology cannot explain the observed increase in galactic densities with distance, and so is withdrawn; big bang cosmology, which anticipates the increase, is advocated for this reason. Explanatory reasoning itself is not in question in such episodes, but is being used to settle questions about the credibility of hypotheses or theories.

If this is a legitimate basis for the shifting fortunes of theory, then the conclusions to which it leads scientists are to be accepted. What to believe, to disbelieve, or to regard with neutrality may be read off the results of scientific evaluation. No *philosophical* theory of science, no argument for a general interpretation of

---

formulated in first-order logic. The presuppositions of applying this theorem to scientific theories prevent it from establishing EE as the thesis about science that I have formulated. (For argument, see Larry Laudan and Jarrett Leplin, "Empirical Equivalence and Underdetermination," *Journal of Philosophy*, 88 (1991):449–472. But the thesis has other than formal motivations in the epistemological writings of antirealist philosophers of science.

4. The view that explanation is wholly pragmatic was rejected in chapter 1. I decline to enter into the details of van Fraassen's grounds for favoring a pragmatic theory of explanation, because the reason that he himself takes to be the most compelling strikes me as simply a mistake. He claims, in *The Scientific Image*, that scientific explanations are not asymmetric, as they would have to be to carry epistemic import. If explainer and explainee can switch roles with changes of context, then abductive inference is circular. I find van Fraassen's own prime example of such a reversal to be, on analysis, fully asymmetric. But the argument I would give essentially repeats that of John Worrall's review of *The Scientific Image*, "An Unreal Image," *British Journal for the Philosophy of Science*, 35 (1984):65–80.

5. A. Fine, "The Natural Ontological Attitude."

scientific theories, is material. Philosophy of science in the role of interpreter and evaluator of the scientific enterprise, and realism in particular, as such a philosophy of science, are superfluous.

This is actually the stance which Fine comes to adopt. His "natural ontological attitude" (NOA), encountered in chapter 1, denies that science in general, especially scientific theorizing in general, is an appropriate object of philosophical scrutiny. This is partly because Fine denies science the status of a "natural kind"; nothing demarcates science, as such, from other forms of inquiry; nothing distinctively scientific generalizes over all the diverse activities and interests so labeled. But primarily, it is because Fine finds nothing for philosophy to contribute at the level of science in general. He thinks that theories "speak for themselves" and that philosophical disquisitions upon theorizing (unless they are confused, metaphysical speculations dependent upon some a priori warrant that is elusive in principle) are not a voice but an echo.

The only proper role for philosophy with respect to science, Fine thinks, is the investigation of conceptual issues raised by the content of specific theories, investigations that belong to the philosophies of specific sciences at specific times, rather than to the philosophy of science in general; and in this role philosophy is continuous with science, rather than a second level of analysis or evaluation. According to NOA, we are entitled to assert whatever science itself comes to assert. There is nothing more to add, and nothing less is compatible with the standards of warrant that license ordinary beliefs. In particular, no questions about what it means for what we assert to be true, about what it is to attribute truth to theories, about what justifies such attributions, about the epistemic status of theoretical entities and their properties—questions which produce philosophical interpretations of science—are proper. Such questions, to the extent that they are not answered as fully as they can ever be by the content of science itself, are systematically misconceived. They have no coherent, warrantable answers, neither realist, pragmatist, instrumentalist answers, nor others. NOA declares a plague on all these houses. It is a neutral, or, as Fine prefers, a "minimalist" position.

But this drastic curtailment of the scope of philosophy arises from *accepting* the legitimacy of explanatory reasoning, of abduction, within science. For Fine does not doubt that science must, and does, use such reasoning to reach the conclusions that NOA endorses. It is not, therefore, the *legitimacy* of abductive inference that is problematic, but the superfluousness of the ends it can serve within philosophy. But ironically—indeed, inconsistently—Fine does dispute the legitimacy of abduction. He deploys the skeptical historical induction to cast doubt on the reliability of abduction within science. And it is not principally superfluousness that he wishes to convict abduction in philosophy of, but circularity. Since, he reasons, abductive methods of theory evaluation are known, via the historical induction, to be unreliable in science—to be as likely as not (in fact, more likely than not) to issue in theories that are later rejected—the attempt to ground a realist view of theories on abductive argumentation is question-begging. Either such argumentation is to be rejected everywhere for its failures in science, or, if it is to be trusted in philosophy despite failing in science, some further argument is necessary to show that it can be trusted in philosophy. And that further argument cannot itself, on pain of circularity, be abductive. Abduction is, at the very least, suspect in the defense of realist philosophy.

Accordingly, a dilemma is created for realism: Either abduction is legitimately used to reach theoretical conclusions in science, in which case realism about science is superfluous and philosophy can add nothing to the specific conclusions that scientists themselves reach; or abduction is suspect within science, in which case realism, if possible at all, must be based, directly or indirectly, on some nonabductive form of argument. An abductive defense of realism, if not pointless, is question-begging. And the irony, lately noted, is that whereas it is the first horn of the dilemma that buttresses Fine's official position, NOA, it is the second horn that he adopts in mounting his attack on realism. For he sees realism typically being defended *as an explanation* of scientific success, without recognition of the need to ground that abductive strategy on anything deeper.

Superfluousness is relatively innocuous; it is vicious circularity that Fine wants realism saddled with. So we have a position, NOA, which would have us regard realism as merely superfluous, and which, for this reason, itself sounds rather realist (although it explicitly denies itself the resources for any such self-characterization). Yet the architect and chief advocate of this position, Fine, disowns realism as logically defective.[6]

More specifically, Fine takes it to be an unacknowledged burden of realism to defend its use of explanatory reasoning. It is incumbent on any philosophical scrutiny of science to evaluate, and if, as in realism, the evaluation be positive, to underwrite, the modes of inference used within science. For the whole point of philosophical scrutiny of science is to determine the epistemic status of scientific conclusions, and this requires evaluation of the reasoning that produces those conclusions. But a philosophical argument does not underwrite a form of inference if it itself relies on that very form of inference for its own defense. This realism, traditionally, has presumed to do.

Fine analogizes the situation to one that faced mathematicians concerned to establish the consistency of a mathematical theory. To appreciate the analogy, we need to distinguish levels of inquiry. The first, or ground, level is that of the theory under scrutiny. The language in which this theory is developed and defended is the *object language*. Then there is the second, or metalevel, at which the scrutiny is conducted. A mathematical proof of the consistency of a mathematical theory would be a metamathematical argument, conducted in a *metalanguage*. Now, it does no good, the mathematician David Hilbert maintained, to use, in one's metamathematical proof of consistency, the very form of inference on which the theorems of the mathematical theory whose consistency is to be established depend. Such a "proof" would be trivial, in that it would be guaranteed to work, but its own validity would be as much in question as that of the first-level proofs that its demonstration of consistency is supposed to underwrite.

---

6. Fine begins "The Natural Ontological Attitude" by proclaiming the death of realism. But many realists find NOA quite congenial. This irony is nicely captured by Laudan's remark: "Realism is dead; long live realism!"

It would reach its conclusion only by presupposing it. Analogously, a defense of realism based on its ability to explain the success of science is possible but trivial, because it presupposes that the ability to explain warrants inferring the truth of the explanation.

I have already betrayed my dubiousness about the viability of NOA, and will have more to say against it. Obviously, if NOA is correct then the project of this book (and those of so many others) is misconceived. Obviously, I cannot be sanguine with superfluousness as the alternative to circularity for realism. But dissatisfaction with NOA is no response to the dilemma Fine creates. This dilemma challenges explanationist realism, and thereby MER, directly. Neither horn of the dilemma must be allowed to survive. I shall respond to the first three challenges and to Fine's dilemma in turn.

The Skeptical Historical Induction

A number of responses to the skeptical induction are logically open, which is probably enough to attract advocates to them at some point. Accordingly, I shall consider some of them without troubling much over whether or where they are represented in extant writings. Most obvious is to make an exception of current science. We need not go back very far in history to find weaker standards of experimental controls and quantitative precision than those in force now. How far back do we go before leaving the science on which it is relevant to induce? Realist claims for science characteristically carry some proviso that the science be "mature." For physics, Newton, or the scientific revolution of the seventeenth century, is a standard threshold; but sometimes, 1905, when confidence in the conceptual basis of physics collapsed and relativity and quantum theory were born, is cited. If the latter choice is defensible, why not the advent of quantum mechanics in the 1930s, when the conceptual and methodological changes were even more fundamental? Starting this late, there is no danger of compiling a record of once-successful but now refuted theories. Quantum mechanics has been tested repeatedly, to incredible levels of accuracy and across a range of applications unimagined at its inception, and never been contradicted. If we put enough contemporary science into the inductive base and demote enough older science to protoscience that can be left out, the skeptical argument collapses.

The trouble is that this procedure is arbitrary. Without a principled criterion for circumscribing the relevant inductive base—for "maturity"—we face a new induction to the likelihood that the threshold will keep shifting, which is as damaging as the original induction. And if a principled criterion is not to require uninterrupted experimental adequacy, which would render its response to the argument circular, it is hard to see how it could disqualify the eighteenth and nineteenth centuries.

An interesting defense is available for making an exception of current science: The skeptical induction itself must make an exception of current science even to get started. The judgment that past theories are false cannot be sustained by looking at past test results in isolation. One must appeal to the implications of the theories that replace past ones. *It is from the perspective of newer theories that older theories may be declared false*; without this perspective we have a multiply interpretable situation, and no means to resolve it. While direct clash with experiment may be enough to preempt a direction of theorizing, developed, successful theories have the resources to withstand such problems, and it is generally up to subsequent theoretical developments to pronounce their fate. But this means that the skeptical induction must assume that current science is correct to obtain the evidence from which to induce that it is incorrect. If the current perspective is incorrect, then how do we know that the theories it replaced are not true?

One possibility is to cite inconsistencies among theories historically discredited. If each of two inconsistent theories was, at one time, successful and both have been rejected, then at least one once-successful theory must be false, whatever the status of current science. One reservation about this answer is that the induction has not been based on inconsistent pairs, but has focused on the *number and variety* of once-successful theories that are no more. Whether it would look as compelling if restricted to cases of inconsistency is questionable. Of course, the inducer might just be assuming that theory succession involves inconsistency, that in replacing one theory by another one is denying the earlier theory. The traditional view, by contrast, is that replacement is not denial but improvement in comprehensiveness and scope of application. While this view is no longer tenable, its failures do not establish that the actual relation among successive theories need amount to logical inconsistency. My own observations about the notion of partial truth suggest otherwise.

A related reservation is that in judging the relations among past theories, we may make assumptions taken from subsequent science. Often, theories that seemed incompatible rivals in historical context are reconciled from a later perspective. For example, rivalries among accounts of absolute motions are obviated by the relativity of motion. To the extent that later science affects our judgments of the relations among historical theories, the problem for the skeptical inducer reappears.

The point may seem legalistic. If we do continually come to judge that our theories are false, one can induce on this fact about us, and infer that, by our own standards, no theory is to be believed. It is not necessary to inquire separately into the appropriateness of our standards to give the induction its punch. But as our present burden is only to fault a criticism, not to defend the view criticized, the point is fair. Its implication is that if we are going to make an exception of current science on the basis of its empirical support, while, at the same time, projecting its failure, we ought to be willing to resurrect once supported theories that subsequent theories defeated.

Let us, then, see what happens if we decree that, whatever science comes to decide, the theoretical posits of successful theories must be real. Instead of getting our view of reality from the latest science alone, we get it from all science that has ever been empirically successful, and complicate it accordingly. Phlogiston exists as well as oxygen, as does the electromagnetic ether. The idea that we are constrained to choose was always a mistake, in that it wrongly presupposed that some single mechanism had to be responsible for a common range of effects.

Rather than regarding light as both particle and wave, or partly each, we can regard it as *sometimes one and sometimes the other*, depending, now, not on what effect we are measuring, but on what theory we are testing. Contrary to our metaphysical presuppositions, no single theory of light need be true of it under all its operations; light can be whatever the theory we are using at the time says it is, when that use issues in predictive success. There need be no single truth about combustion; we can have it both ways without inconsistency by diversifying our metaphysics.

What happens is disastrous, and I display this view only to preempt it. It offers a pluralism that may appeal to the same egalitarian sensibilities as ethical pluralism and cultural relativism; the realism it protects is worth no more than the morality or culture that issues from an indiscriminate celebration of differences. Unless the synchrony of theory and world is a monumental coincidence, on the order of predestined parallelism in the philosophy of mind, the realism left is no better than George Berkeley's. To suppose that the act of theorizing, whenever it leads to successful prediction, determines the ontology responsible for what is theorized about is to create an idealism of the unobserved.

And it does not stop there. For how is the consistency of alternative theoretical posits to be maintained when different scientists simultaneously apply different theories to the same phenomenon to produce the same correct prediction? Now phlogiston will have to be released and oxygen absorbed both in a single case of combustion. Evidently, the single case of combustion must divide into one for each scientist, and we get idealism all the way down. Now all experimental controls on which theories get to generate experience collapse, and, with them, the responsiveness of this line of reasoning to the original argument.

A rather different, more natural response to the skeptical induction is to discover a sudden rectitude with respect to standards of inductive inference. Although rough-and-ready induction is the stuff of daily life, we might understandably demur at so loosely anecdotal an inference as that to the falsity of all possible theories from the fact that some impressive ones have proved inadequate. No one has counted theories to see how many such failures form an inductive basis, let alone offered any handle on the scope of what that basis is supposed to disqualify. We have no probabilities, no measure of the level of confidence in the conclusion that the evidence warrants. The evidence that disqualifies past theories is alleged to be so powerful as to refute not only them but also all theories ever to be entertained, whether or not it bears any direct relevance to their subject matter.

Such reasoning looks suspicious; it ignores potential complications of the sort that produce familiar fallacies of induction. What if one induced that all athletic records are bound to be broken from the fact that many records that once looked so daunting eventually were broken? That is, what if such a conclusion were drawn without regard to the nature of individual records and the types of performance they measure? Obvious theoretical limits are thereby neglected. No one will run a zero-minute mile, and given the finite resolution of chronometers, some record at some point must stand. Each record set *reduces* the likelihood that it, in turn, will be broken, not only by reducing the range within which subsequent records must be made, but also by being harder to break.

Suppose that theorizing is like this. Theories, as I have urged, are defeated not merely by adverse observations, which are commonplace in the best of times, but by better theories. And the better a theory is—the more that it explains and the better the explanations it gives—the greater is the onus on a theory that would be better still. Maybe the world sets limits to how good a theory can be by subsuming itself completely under some potential, final theory. Maybe the history of theory change has brought us closer to a final theory. The possibility cannot be dismissed a priori. Indeed, I have noted that quantum theory anticipates limits to sizes and times at which uncertainty principles block further theorizing, and big bang cosmology imposes a similar limitation on cosmological theories that fall under it. Given this picture, the historical induction fallaciously neglects evidence whose inclusion in its inductive base would block the skeptical inference from it. If current theories owe their relative impressiveness to a process that requires improving on, and thereby rejecting, earlier theories, then the fact that earlier theories were rejected is not evidence that current ones will be. That would be like citing one's longevity as evidence of one's immortality.

This argument creates a kind of dilemma for the skeptical inducer. He wants past failures to ordain future ones, but the process of theory change does not allow him the natural implication that had past theories *not* failed—had there not been those transitions—then science would be better off than he induces it to be. To the contrary, science looks far better off for its past failures, because they have been the route to improvement. Indeed, part of the confidence we have in viable theories depends on how they contrast with theories that have failed; failures are needed to recognize success. But one cannot have it both ways; past failures cannot be both the route to improvement and the basis for skepticism. The dilemma is that in addition to getting skepticism from past failures, we also seem to get it from the supposition that past theories never get replaced. For then, our understanding of the world never improves; its limits and errors are never recognized. We avoid skepticism on this horn of the dilemma only by imagining that we get everything completely right the first time around, a picture without basis in our experience of knowledge and learning. Actually, it is characteristic of epistemological skepticism, in general, to make immediate, infallible knowledge the only kind permissible, to suppose that once the possibility of error gets a foothold, as it does in any ampliative, self-corrective epistemic process, this foothold can never be surmounted. The inevitable conclusion under this limitation is that there is no empirical knowledge at all.

But for both horns to issue in skepticism is surely an embarrassment. It means, minimally, that the fate of past theorizing is incidental; what really matters to the skeptical view of current science is not that past theories failed, *but that there were any past theories at all*. The real problem is that history produces alternatives. It is some inability to choose that turns out to be the real threat to knowledge. Perhaps, then, the skeptical induction reduces to the problem of underdetermination, and requires no response on its own terms.

Or one may prefer a rhetorical response, in the manner of the counterinductions that inductive fallacies encourage. For example, as secure as the judgment that past theories have failed is the judgment that they have improved on their predecessors.[7] These judgments have equal historical basis. So the claim that the second horn of the dilemma also leads to skepticism, issuing as it does from a progressive reading of the process of theory replacement, has as firm an inductive basis as the claim that the first horn leads to skepticism. We are as warranted in concluding that our understanding is improving as we are, at any point, in concluding that it is yet improvable. But the realism that the skeptical induction opposes does not suppose us ever to be in a position to proclaim our understanding complete, nor does it suppose that our understanding will ever *be* complete, proclaimably so or not. Requiring only epistemic progress (or, in MER, only the possibility of empirical warrant for claiming epistemic progress), realism is as much inducible as skepticism.

I am not going to prolong this line of criticism, because I do not expect it ultimately to satisfy. One theme running through fallacious examples like the one about athletic records, which are supposed to embarrass inductive generalization, is the absence of an appropriate explanatory underpinning, which chapter 5 argued is necessary to support generalization. The skeptical inducer is in a stronger position. The likelihood of failure is readily understandable against an assumed background of unlimited theoretical options that observation can only partially discriminate. This background may be independently challenged by arguments against underdetermination. But when all the logical misgivings are voiced and the embarrassing analogies played out, some bite will remain to the claim that our conditions for investing credence have clearly been met by theories that do not deserve it. To argue that this is compatible with epistemic progress, and that the route to progress would not be expected to proceed without error, is ultimately question-begging if progress has not itself been demonstrated. What we are judging is a method of judging, and such a method must ultimately answer for its track record. It must be judged by its reliability; naturalism leaves no alternative. Not knowing, independently of what our best theories tell us, whether the world is such as to place limits on theorizing analogous to those that defeat inductions about athletic records, we are able to offer but speculation against historical evidence that the method is too unreliable to be trusted to produce true theories.

I therefore propose to confront the skeptical induction directly, by declaring that *where past theories have met the standards I have imposed for warranting theoretical belief, their eventual failure is not total failure; those of their theoretical mechanisms implicated in achieving that warrant are recoverable from current theory.* I stipulate that if theoretical mechanisms used to achieve prolonged records of novel success uncompromised by failure have regularly been obviated by subsequent theory, so that, from the perspective of current science, nothing of them survives, then the antirealist, inductive skeptic wins.

This is a significant test. There are many theoretical mechanisms of past sci-

---

7. This point was broached in chapter 1 and revisited in chapter 5. Not wishing to follow certain other authors in using it without attribution, I will cite its original source. It is myself, in "Truth and Scientific Progress," *Studies in History and Philosophy of Science*, (1981):269–293. A revised version appeared in my edited volume, *Scientific Realism*.

ence—the crystalline spheres of ancient astronomy, phlogiston, the electromagnetic ether—of which nothing survives, and there is a general criterion of total demise available in the preemption by subsequent theory of the very role that a past theoretical mechanism was designed to play. I submit MER to this test, and will not exploit the problematic logical routes I have charted to protect it. My claim is that MER passes; if realism generically is discredited by the skeptical induction, MER is not. This claim discharges my obligation to naturalism. More specifically, it discharges my commitment to show that my case for realism, while explanationist, carries defeasible empirical consequences.

To make good on this claim would require, ideally, an examination of historical cases in which theoretical beliefs have been justified, by my lights, and their comparison with current theory. What I would hope to show is that although theory change is not uniformly or fully cumulative, and theories are not, in general, special cases of their successors, there is enough cumulativity where it counts; justified beliefs are at least partially sustainable within the deeper perspective of later theory. The trouble is that history offers precious little in the way of empirical conditions that meet the exacting standards of warrant that I have imposed. MER does not require it *ever* to have supplied them. The burden of my own examples of novel success was to explicate the concept of novelty, not to advocate specific theoretical beliefs. These examples, for the most part, are both too limited, in not providing a sufficiently extended record of success to establish warrant, and too contemporary to urge against the inductive skeptic.

On the other hand, the favorite examples of the skeptic—the rotating spheres model of the universe; phlogistic chemistry; the caloric theory of heat; the electromagnetic ether—cannot claim novel success in my sense. Their failure is no threat to my argument. There may not be enough novel success to make good on my claim, but then neither is there enough to support a skeptical induction against MER. It would be enough for my purposes to leave it at that, to offer my claim as a challenge to antirealism, rather than to try to prove it by cases. I can do better, however. I can offer a few points in connection with particular cases as motivation.

Fresnel's theory of diffraction is, arguably, old enough and successful enough to serve as a test case between myself and the skeptic. That this case conforms to my claim follows from the very able account of it recently given by Philip Kitcher in his ambitious work *The Advancement of Science*.[8] I cannot, and need not, improve on Kitcher's account, and will not repeat it. But if I am to invoke it, I do need to defend it against a predictable criticism.

Kitcher maintains that some of Fresnel's uses of the concept of the ether amount to background interpretation, inessential to his use of his theory to obtain

---

8. Philip Kitcher, *The Advancement of Science*, chap. 5, sec. 4. Kitcher does not himself engage the concept of novelty, nor attempt any general analysis of the conditions that warrant theoretical belief. He operates, instead, with an eliminativist conception of evidence. His defense of realism is therefore much different, and, I contend, weaker, than my own. For criticism of his strategy, see my review article in *Philosophy of Science*, 64 (1994):666–671. (Erratum: p. 667, par. 3, l. 2: for "of whose," read "whom.")

specific diffraction patterns. On other occasions, he uses the concept ineliminably in applying the theory. Only where the concept is inessential do his applications stand up to empirical test. On these occasions, Fresnel is to be read as committed to something far less than the full ether concept, to a more modest and general notion of light as a wave phenomenon that survives in modern science. This division of uses nicely explains how Fresnel's success can fit into a cumulativist picture of theory change. The question is whether the division can be made in a principled way that does not beg the question of cumulativity. Is it not, instead, an arbitrary expedient that reads retroactively into history just the distinction that realism needs?

I suggest that the ability to divide uses as Kitcher proposes, in context and in a principled way, has an ample basis in the general theory of reference. Perhaps in the case of proper names, but certainly in the case of descriptions, the success of reference does not require the referring expression to be true of its referent. If the individual I identify as "the woman drinking champagne" is actually drinking a white wine made by the *méthode champagnoise* in the Loire Valley, not in Champagne, then the description I use, though it refers to her, is false of her. Evidently, something other than accuracy of the description accounts for the establishment of a referential relation between it and the drinker. What makes it the case that this particular individual, rather than someone else or no one, is the referent must be something about the occasion of the description's use, rather than something general about the description or the referent. Ostension is the most straightforward possibility, and there were many circumstances in which Fresnel's experiential relationship to the phenomena he was analyzing was sufficiently unambiguous not to require the accuracy of his theoretical description for the success of the reference. It was because it was light that his ether-theoretic description referred to, and because light propagates as waves in the diffraction phenomena he was investigating, that his predictions were successful, although his theoretical account was erroneous. Similarly, the phlogistonians had lots of predictive success despite their use of the concept "dephlogisticated air" in analyzing it, when it was actually oxygen-enriched air that was before them.

The implication is that although it is correct to deny that the ether or phlogiston exist, it is not, in general, correct to deny that 'ether' or 'phlogiston' refer. These terms did refer on many occasions, although what they referred to on those occasions was not ether or phlogiston.[9] Of course, it would be convenient to have a general account of reference that would enable questions like "Does 'ether' refer?" and "What must be true of something in order that 'ether' refer to it?" to have general answers that depend only on the term 'ether' itself and the theoretical context of its introduction. This convenience is recoverable by introducing a distinction. Kitcher's distinction is between the term itself as a type and its tokens, its utterances or inscriptions on particular occasions. It is the tokens whose refer-

---

9. Whether, having distinguished 'phlogiston' from its tokens, or its reference from that of its users, one can then deny that 'phlogiston' refers is a question that raises further complications. Certainly 'phlogiston' is not true of anything, but referring and being true of are not the same.

ence gets diversified, relative to features of the occasions that produce them. A different possibility, older than Kitcher's, is to distinguish "semantic reference" from "speaker's reference," or reference from denotation.[10] We can then continue to speak, in general, of the reference of a term, or of semantic reference, and handle the phenomenon of referential diversification under another label. Plainly, there is such a distinction, as one's ability to refer to the drinker though mistaken about her beverage attests. And this indicates that the discriminations that Kitcher relies on to extricate rejected theoretical entities from the success of predictions made when using them need not be arbitrary or retroactively imposed. The context makes them evident.

Only when reference depends on the theoretical account itself, rather than on features of the context of its use, does my response to the skeptical induction require the account to survive in subsequent theory. Realism allows empirical warrant to accrue to a theory whose key terms do not refer (to speak in the convenient, general way) if that warrant is achieved in ways that do not depend upon using false theoretical conceptions on whose basis the terms were introduced. Not just any of a rejected theory's theoretical mechanisms that clash with current science can be regarded as mere adventitia, inessential to that theory's empirical warrant, but some of them can be, at least on some occasions of their use. Those that cannot, and that must therefore be recoverable from current science, are those that were necessary for achieving the rejected theory's empirical warrant. The cumulativity that it is incumbent on MER to discern in theory change is the retention, in the form of acquiescence if not outright endorsement, of such portions of rejected theories as were regularly relied upon to achieve for those theories a substantial record of novel success uncompromised by failure.

This much cumulativity it is not implausible to expect. We reject the ether but continue to regard light as, in part, a wavelike phenomenon, preferring the conceptual difficulty of endowing space itself with the ability to transmit disturbances, to abandoning altogether the wave picture. Much of Fresnel's conception was right, by our lights, enough of it to account for the novel success of his theory. Newtonian theory can claim much novel success—the prediction of new planets (not their elements of orbit insofar as these could have been used interchangeably with those of Mars to generate Kepler's laws, but their existence),[11] the variation in gravitational acceleration with distance from the gravitational source. And while Newtonian gravity has been superseded, current physics maintains the universality of gravitation and Newton's key theoretical insight of identifying the mechanism responsible for the phenomenon of weight on earth with

10. These distinctions originate in Keith Donnellan, "Reference and Definite Descriptions," *Philosophical Review*, 75 (1966):281–304; and in Saul Kripke, "Speaker's Reference and Semantic Reference", in Peter French, Theodore Uehling, and Howard Wettstein, eds., *Contemporary Perspectives in the Philosophy of Language* (Minneapolis: University of Minnesota Press, 1977).

11. The additional outer planets could have been discovered telescopically (perhaps in a systematic survey comparing relative positions of photographic images over time intervals), and then Newton's law used to compute their orbits, rather than the other way around. In this case, on my view, the predicted orbits would not have been novel for Newtonian theory.

that responsible for the motions of the planets. Newton's discovery that free fall, pendular motion, lunar acceleration, the acceleration of the moons of Jupiter, and the acceleration of the planets instantiate a single theoretical mechanism identified as universal gravitation is unqualifiedly affirmed in contemporary physics. We vacillate in representing this mechanism in terms of the concept of attraction. But the Newtonian law of gravitation, to which we attribute novel successes, was deduced from Newton's mechanical laws, and laws of Kepler and Galileo, which did not depend for their rationale on a concept of attraction. It is as precipitous to reject all claims to cumulativity where a once-invoked theoretical mechanism has been abandoned, as it would be to uphold all theoretical mechanisms ever to figure in successful prediction. The possibility of a non-question-begging criterion of what to uphold that produces the cumulativity MER needs, though complex in application, is plausible enough.

Moreover, there are often powerful reasons independent of empirical success for projecting the retention of a successful theoretical mechanism. The entire discussion of the skeptical induction has proceeded under the idealized and misleading assumption that theories can be divided into discrete, manageable units that survive or fail in confrontation with empirical evidence, independently of the wider theoretical contexts in which their basic hypotheses arise. Many philosophers of scientific change—Kuhn, Lakatos, and the arch historical skeptic himself, Laudan—locate theories within larger traditions that they hold to be relatively resistant to empirical refutation. Yet when it comes to inducing antirealism from history, it has been supposed that the failure of a specific theory is sufficient to refute the theoretical hypotheses that that theory's successes had been taken to support. I shall use one further example to illustrate the challenge that sensitivity to a wider theoretical context poses to this idealized picture.

I passed over my own examples of novel success as too contemporary to test the strength of the induction. But in one case, big bang cosmology, there is already the prospect of theory replacement. Despite its novel successes, big bang cosmology may be supplanted by a theory proposed in 1983 by Jim Hartle and Stephen Hawking.[12] According to big bang cosmology, the universe begins with infinite density and infinite curvature. These conditions constitute a singularity at which the laws of physics break down. The "no-boundary" proposal of Hartle and Hawking avoids this limitation on the scope of physical law by applying the sum-over-histories interpretation of quantum mechanics (the Feynman path-integral formulation) to the universe as a whole. Their technical innovation is to do the summation in complex time. The path integrals form a space all of whose dimensions are spacelike, as opposed to the usual formulation, in which each path has a direction in time.

The result is that the wave function of the universe is finite and nonzero at the singularity, just as is the wave function of the electron in a hydrogen atom. As the electron can "pass through" the proton—has a nonzero probability of occu-

12. James B. Hartle and Stephen W. Hawking, "Wave Function of the Universe," *Physical Review,* D28 (1983):2960–2975.

pying the position of the proton—so can the wave function of the universe pass through the singularity to the other side. But because time is complex, the "other side" always represents the forward direction in time. Thus, although finite, time is unbounded. The universe has no earliest moment in complex time. Hawking analogizes the earliest moment of the universe in real time to the North Pole of the finite, unbounded surface of the Earth.[13] Nothing physical distinguishes the North Pole from other (neighboring) locations, yet all directions from it are south. Similarly, physical laws apply at the real-time singularity as at all other times, yet at the singularity all directions in time are forward. The analogy works because using complex time makes time geometrically like space; it can be curved, because it makes sense to consider positions in time along an axis orthogonal to the real-time axis.

The Hartle-Hawking theory is tentative and problematic. Its restriction to paths that lack temporal directionality appears arbitrary—motivated by the desire to achieve an outcome free of singularities, rather than derived from anything more fundamental. It is not clear how to restrict the paths in a suitably principled way.[14] But I will suspend such concerns and entertain the possibility that the theory proves successful. The question is whether it obviates big bang cosmology. My answer is that it does not, because big bang cosmology is embedded within the larger theoretical program of general relativity, which the no-boundary proposal does not challenge. It is that very program which indicates that in the early universe conditions prevail under which quantum effects become dominant, and which thereby makes the proposal possible. The point of the proposal is to extend the laws of physics to all times, to obviate the need to supplement them with boundary conditions. The point is to avoid singularities, not to eliminate the physical idea of origin. That there must be a singularity at the beginning of the universe in real time is a consequence of the history of the universe in complex time. And real time, in the mathematical sense of 'real,' does not become unreal, in the ordinary sense, just by becoming embedded in a wider theory.

On the other hand, should the new theory be interpreted as eliminating the big bang from the history of the universe, the epistemic importance of the novel success of the big bang theory could still be protected by applying my criterion for the retention of successful theory. For an originating event at infinite density and curvature is not necessary to obtain the novel results I identified. Expansion from a state of high energy and density is sufficient. The no-boundary proposal retains this evolutionary portion of big bang cosmology by providing for a transition from Euclidean to Lorentzian space-time with a timelike dimension.

Such discussion of examples is the only way to engage the empirical issue that divides me from proponents of the skeptical historical induction. My discussion does not prove that all theoretical mechanisms used essentially in achieving sustained records of novel success, uncompromised by failure, are recoverable from

13. Stephen Hawking, "The Edge of Space-Time," *New Scientist*, 16 August, 1984, pp. 10–14.

14. For discussion of the problems and prospects of the theory, see Don Page, "The Hartle-Hawking Proposal for the Quantum State of the Universe," in Leplin, *The Creation of Ideas in Physics*.

current theory. But it does shift the burden of argument back onto those who take the induction to defeat realism, and more than that, it rebuts reasons for reticence as to cumulativity through theory change and motivates the expectation that as much cumulativity as MER requires will be discernible.

There is, however, one possible suspicion about my strategy that this modesty may not allay. I disarm the great majority of examples of rejected but once-successful theories as not having achieved success that is novel in my sense. Only where success was novel am I under any onus to display cumulativity. But there is an intermediate position. Success may fail to be novel for any of a number of reasons, pertaining either to independence or uniqueness. Are my historical responsibilities limited to novelty as such, or might they extend to cases where novelty was achieved "in part," where it would have been achieved but for reasons that, with respect to theory retention, appear incidental? Might a result that does not meet the full letter of my analysis yet approximate it well enough to pose a historical challenge? If so, one might suspect that the stringency of my analysis has been exploited inappropriately to evade responsibility for potential counterexamples. Where, specifically, I anticipate this suspicion arising is with respect to a theory whose successes would have been novel if only an alternative theory predicting them had been delayed or more quickly disconfirmed. Should such a theory be subject to my cumulativist commitments? The suspicion I anticipate, in short, is that my historical responsibilities should be fixed by the independence condition alone.

I do not assign partial truth to a theory whose predictive success meets only the independence condition, because such a theory is not the only basis for the prediction. My argument for realism does not require that truth attributions be justified in all cases of predictive success, and cases that fail to meet the uniqueness condition are unnecessarily problematic for my purposes. But how, in such cases, *do* I explain the theory's predictive success? If truth is not the right explanation in such cases, then why should it be the right explanation in cases of novelty? It is evidently the independence condition alone that establishes the *need* to invoke truth as explanation of success. The point of uniqueness is to remove alternative predictions of the same result, not alternative explanations of a theory's success in predicting the result. So if, though needed, truth is not to be invoked in the one case, then perhaps it should not be invoked in the other. And if truth *is* to be invoked on the basis of the independence condition alone, then are cases that meet this condition not fair candidates for retention through theory change, whether or not they also satisfy uniqueness?

In fact, I do not explain how a theory manages to achieve predictive successes that would be novel but for failure to satisfy the uniqueness condition. Truth may well be the explanation in such cases, especially if the alternative theory that defeats the uniqueness condition does not itself satisfy the independence condition. Should both theories satisfy independence, the explanation may be the truth of some common hypothesis responsible for each theory's prediction. I am certainly inclined to think so. It is just that my argument does not require me to make such claims. And my cumulativist commitments cannot be forced on theories to which I am not obliged to attribute truth in the first place.

I do not see, however, that extending my commitments in the way suggested would make much historical difference. It is not as though I have had to rely on the uniqueness condition to defuse examples commonly proposed for the skeptical induction; the independence condition alone disqualifies them. There is a general reason not to fear a lowering of my historical defenses. The best examples for the induction are rejected theories that not merely claimed some empirical success, but also were once accepted. It is the fact that we have been wrong in accepting theories that is supposed to undermine standards of acceptance. But where rival theories successfully predict the same results and are not otherwise impeachable—conditions that violate uniqueness—scientists are unlikely to accept any of them. This is a feature of the scientific practice of according novelty special epistemic significance that my analysis of novelty recovers. It is not to be expected, therefore, that imposing the uniqueness condition will disqualify cases to which my cumulativist commitment would otherwise apply. This commitment is incurred only on behalf of theories whose empirical success, in comparison with competitors, is sufficiently large, sustained, and unblemished to enlist the credence of the scientific community. Although that credence is sometimes invested prematurely by my lights, under conditions that do not require truth for an explanation of success, it is not invested arbitrarily among competing alternatives. And if a theory was never believed in the first place, it is hardly encumbent on me to provide for its retention. If it is reasonable to ask whether once-accepted theories whose success was not novel should nevertheless be subject to my cumulativist commitments, it is surely unreasonable to demand that these commitments extend to theories that were never accepted. And for theories that do get accepted, we would expect to find the uniqueness condition satisfied.

So what do I say to the suggestion that, on explanationist grounds, the invocation of truth must not depend on uniqueness? I invoke truth where it is needed, provided there is no independent obstacle to doing so. I do not see how this obligates me to invoke it wherever it is needed. Where it is needed but unavailing, as in the absence of uniqueness, I admit uncertainty. That I can, as well as need, do no better than chance as an explanation for predictive success in some conceivable situations is no reason not to do better where I can.

Empirical Equivalence and Underdetermination

EE (by which, recall, I mean the *thesis*, not the condition, of empirical equivalence) may be interpreted as requiring either the possibility or the actuality of empirically equivalent rivals to any given theory. I need not contest the former, weaker version, because the possibility that a belief is wrong does not defeat its justification. Certainly the falsity of any number of unproblematic beliefs rooted in ordinary perception is strictly compatible with the evidence that justifies them. Justifying evidence does not, in general, preclude all scenarios under which a rival belief is true. To undercut justification, one needs, minimally, some evidence, some grounds for suspicion, that such a scenario is in play. As there can

be no motive for demanding a higher standard of justification for theoretical beliefs than for ordinary perceptual beliefs, the mere possibility that some unknown
rival theory will explain the very phenomena that are taken to support a given
theory does not threaten realism.

This is not to concede EE in its weaker form. Of course, the logical possibility
that rival theories will emerge can be conceded. But the more ambitious claim
that resources invariably exit to generate them is not easily established. I know of
no argument for the weaker form of EE that does not proceed via the stronger
form, and I find arguments for the latter form highly suspect. Such arguments
are constructive in some way; they proceed by example, displaying purported
rivals in particular cases, or algorithmically, showing how, from an arbitrary theory, an equivalent rival may be obtained. As EE is widely held to threaten realism,
I shall offer some objections to these strategies. I shall also consider the alleged
connection between EE and UD, for it is the latter thesis that most directly challenges MER.

The evaluation of claims to empirical equivalence in particular examples can
require extensive argumentation. I shall not examine examples here, both because
I have tried to do this elsewhere[15] and because to attempt to block examples is
to play the antirealist's game. New examples of purported empirical equivalence
may always be proposed, and to assume the burden of disarming all of them is,
in effect, to oppose the weaker thesis, which is unnecessary. It may be stipulated,
for the sake of argument, that there are individual cases of empirically equivalent
theories.[16] Unless one attempts a sweeping generalization from them, they are no
threat to a realism that, like MER, does not require the possibility of warrant for
theories generally. Moreover, the equivalences stipulated are subject to a significant qualification: It is possible for them to break down under further investigation. All claims to empirical equivalence in particular cases must be relativized

---

15. In "Empirical Equivalence and Underdetermination," Laudan and I construct a generic example to show that EE cannot be founded on the indeterminacy of absolute motions. The examples of
purported empirical equivalence that are used to attack realism typically exploit the relativity of motion in mechanics. There are more sophisticated examples in which cosmological models are made
to diverge in their empirical commitments only outside the range of events whose accessibility is
allowed by relativistic limits on causal connectability (John Earman, Clark Glymour, and John Stachel, eds., *Foundations of Space-Time Theories* [Minneapolis: University of Minnesota Press, 1977]).
The point to make about such examples is, of course, that the equivalences they exhibit have physical
presuppositions that future observations could challenge. The strongest form of empirical equivalence
that any such example can establish is not equivalence in principle but equivalence to the best of our
knowledge.

Nor does mathematical equivalence overcome this limitation. Rival physical theories that are
mathematically equivalent must interpret the mathematics differently. In the absence of a general
proof that all possible empirical results are assimilable under each interpretation, it is possible that
some new result will be predictable under one interpretation but not the other.

16. See John Earman, "Underdetermination, Realism, and Reason," in Peter French, Theodore
Uehling, and Howard Wettstein, eds., *Midwest Studies in Philosophy*, vol. 18 (Notre Dame: University
of Notre Dame Press, 1993). Although this paper purports to be a criticism of Laudan and me, it is
difficult to identify any specific conclusion it reaches that is incompatible with our position.

to the state of our knowledge; they have physical presuppositions that need not survive further theory change, nor change in the scope of what is observationally accessible by technological means. Thus they need not instantiate EE.

In effect, any claim of empirical equivalence is a defeasible inference from current evidence. It may be defeated by new evidence that refutes background theories on which the equivalence is based. But the main source of the limitation is that the very identification of a class of empirical phenomena as those to which a theory is committed by implication is defeasible. Logical relations between theory and evidence are characteristically mediated by a range of assumptions—initial conditions, background constraints, and collateral hypotheses of a theoretical nature. Although these assumptions are usually more secure than the theory they are invoked to test, they are open to revisions that can affect the status of an empirical phenomenon as one to which the theory is committed. As a result, a finding of empirical equivalence among theories must be indexed to the information available for use in obtaining empirical consequences from these theories. With additions to, or deletions from, the body of available auxiliary information, the empirical implications of the theories may diverge.

The involvement of auxiliary information in judging the empirical equivalence of theories introduces a further complication which has not been recognized, and which undercuts the general importance of EE, beyond its instantiation by actual cases of empirical equivalence. The epistemic interest of EE lies in its connection to UD. The point of arguing for EE is to get to UD; theories that do not differ observationally cannot, it is supposed, be differentially supported. But then no such theory can be individually warranted. The situation is rather like a criminal investigation in which the evidence shows that either the butler or the maid is guilty but does not indicate which. Then no one can be convicted. If empirical equivalence is rampant, it would seem that we cannot be realists because we would never know what to be realists about. If UD were not in the offing, EE would lose its punch. A multiplicity of theoretical options is epistemically innocuous if, contrary to UD, the evidence is capable of establishing one of them to the exclusion of the others.

Now, this inference from EE to UD is problematic in its own right. That theories have the same observational consequences does not imply that the evidential weight of these consequences bears on them equally. Nor does it imply that all evidence bears on them equally, absent an independent argument that a theory's observational consequences exhaust the evidence that bears on it. These are known lines of criticism.[17] To them I would add that a result novel for one theory may be presupposed by the other. The difference between novelty and mere status as a member of a theory's consequence class is an evidential difference, according to my argument for realism. If this is correct, then the evidential weights for different theories of a common consequence may differ. But the point I wish to make here is independent of these criticisms. What has not been recognized is that *the truth of UD would prevent the determination that theories are*

17. See "Empirical Equivalence and Underdetermination."

empirically equivalent in the first place. Because theories characteristically issue in observationally attestable predictions only in conjunction with further, presupposed background theory, what observational consequences a theory has is relative to what other theories we are willing to presuppose. As different presuppositions may yield different consequences, the judgment that theories have the same observational consequences—that they are empirically equivalent—depends on somehow fixing the range of further theory available for presupposition. And this UD ultimately disallows.

Of course, there is an easy way to achieve empirical equivalence while respecting the role of auxiliary theory in prediction. If we place no constraints at all on auxiliaries, everything becomes predictable, and EE holds trivially. I take it that the truth of EE is not to be purchased at the expense of such triviality. It is a thesis invested by its proponents with great epistemological moment, as its role in generating UD indicates. UD is not intended to deny that theories admit of evidential support; it denies that such support can be definitive in the sense of (even defeasibly) warranting belief. A UD obtained from a trivialized EE permits no discriminations whatever from evidence, and amounts to a sweeping skepticism that it does not take realism to oppose. Complete incontinence in our willingness to deploy auxiliaries would produce a science that can be neither explanatory nor internally coherent. To carry epistemic interest, EE must be accompanied by some standard for the admissibility of auxiliaries.

That standard, I submit, must itself be epistemic. Admissible auxiliaries are those independently warranted by empirical evidence. Unless auxiliaries are *better supported* than the theory they are used to obtain predictions from, those predictions cannot be used to test the theory. The significance of their success or failure would be indeterminate as between the theory and the auxiliaries. The result would be a holism that enlarges the possible units of empirical evaluation, and prevents epistemic support from accruing to theories directly. Such is the upshot of the classic theses of Duhem, who stressed the ineliminability of auxiliaries from prediction, and of Quine, who claimed the availability of alternative auxiliaries capable of protecting any theory from adverse evidence.[18] But if no theory can be warranted, as UD maintains, then the epistemic standard for the admissibility of auxiliaries cannot be met. In particular, it will not be possible for the theoretical auxiliaries needed to test a theory to be better supported than the theory, in advance of testing it. For they will be subject to the same immunity to probative evidence as afflicts theory in general. And this means that it will not be possible to circumscribe the range of phenomena that qualify as the theory's empirical commitments. *There will be no fact of the matter as to what the empirical consequences of any theory are.* But then there can be no fact of the matter as to whether or not the empirical consequences of any two theories are the same. EE cannot be used to obtain UD, because if UD is true EE is undecidable. If UD is to be believed, it must be given some independent basis, such as the skeptical holism which the theses of Duhem and Quine produce.

---

18. See Leplin, "Contextual Falsification and Scientific Methodology."

In short, the very identification of empirical consequences *as consequences* presupposes, in defiance of UD, that we have the means to decide whether prospective theoretical auxiliaries are epistemically warranted. We must concede that observational evidence unambiguously identifies some theoretical claims as uniquely or preferentially warranted even to raise the issue of EE. The question of whether theories are empirically equivalent must be the question of whether, in conjunction with auxiliaries that *we have already found reason to believe,* the theories have the same observational consequences. If the indecisiveness of evidence were ubiquitous, as UD proclaims, this question could not be formulated.

I take this argument to be definitive against the use of EE to oppose MER, but there are details to consider. It is possible to respond to the argument by resorting to some system of constraints on the admissibility of theoretical auxiliaries short of evidential warrant. Could one not, for example, take *conformity* with observation as a criterion of eligibility, where such conformity is not supposed to constitute warrant because EE ensures that alternative hypotheses share it equally? Given a number of empirically equivalent hypotheses in conformity with observation, choose *any* to serve as auxiliaries since the choice among them cannot affect what is thereby predicted from a given theory.

But now the question arises of what it is for a hypothesis to be in "conformity" with observation, where this relation is not to confer epistemic warrant. If the proposal is simply to use as auxiliaries any hypotheses that observations have not refuted, then Duhemian holism tells us that conformity effects no constraint at all. Observations are incapable of refuting any theoretical proposition absent the mediation of auxiliaries. We would require a principle of selection for further auxiliaries to determine which theoretical propositions are to be admitted as auxiliaries in the first place. The problem of constraining auxiliaries has simply been pushed back a stage, initiating an infinite regress.

Is conformity, then, to be a more robust relation, involving some semantic connection between hypothesis and observation that enables the hypothesis to figure usefully in generating predictions? In this case, to constrain the choice of auxiliaries requires stipulating which are the intended predictions whose derivation from a given theory conforming hypotheses are to facilitate. For different choices of predictions will, in general, require different auxiliaries. Without specifying what is to be obtained as an observational consequence of the given theory, there would be no basis for a preference among prospective auxiliaries.

Note that EE cannot be deployed to evade this result. Applying EE to the theoretical propositions on offer to serve as auxiliaries to a given theory requires every prospective auxiliary to be replaceable by a rival without disturbing the resulting class of observational consequences of the given theory. This condition does not, in general, render the choice of auxiliaries neutral with respect to the class of observational consequences, because the choice of auxiliaries is not, in general, a choice among empirically equivalent propositions. Consider two pairs of prospective auxiliaries, all four conforming to observation, such that empirical equivalence obtains within each pair but not between pairs (figure 6-1). EE is satisfied with respect to all the prospective auxiliaries; the choice among them is not constrained by the requirement of conformity with observation; and yet the

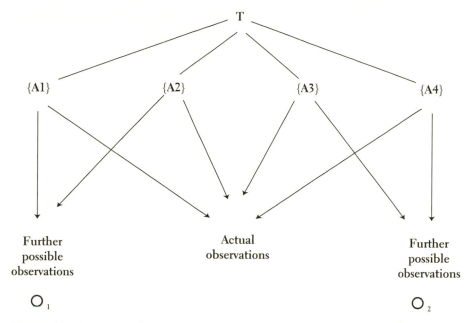

Figure 6-1   {A1}–{A4} all conform to observation; all have equivalents and are, to that extent, underdetermined epistemically. Nevertheless, the choice among them affects the observational-consequence class of T.

observational commitments of the given theory vary with the choice between pairs. One has only to allow for divergence of predictions with respect to possible future observations. Unless the predictions to be derivable are specified, there is no criterion for the choice between pairs. Accordingly, there is no constraint on the eligibility of auxiliaries in general.

But this brings us full circle: We must specify in advance what the observational-consequence classes of theories are to be in order to obtain a criterion for the selection of auxiliaries used to obtain them. The circle is certainly vicious; there can hardly be a principled basis for delimiting predictions independently of the theory and auxiliaries from which they are derived. We would be in the position of establishing EE by fiat, by stipulating the consequence classes and using them to select the admissible auxiliaries. In effect, a dilemma has been created: Either the choice of auxiliaries is unconstrained, different choices yielding different observational-consequence classes for theories; or the status of permissible auxiliary is relativized to the specification of observational consequences, which must now proceed independently of the theory whose consequences they are to be. Both horns of this dilemma vitiate the presumption that theories possess empirically equivalent rivals. All of this is inevitable if, with UD, we deny recourse to a standard of epistemic warrant for auxiliaries. UD renders the connection of theory with observation irremediably arbitrary. It thereby vitiates EE.

To summarize: Unless all theories are to conform to all possible observations, there must be a constraint on what we are allowed to use to effect the connection. We cannot use empirical equivalence itself as a constraint, limiting auxiliaries to any within a class of empirically equivalent hypotheses that conform to actual observations, because there are any number of such classes diverging in their predictions for future observations. Any attempt to make a principled selection among such classes, if it does not beg the question by presupposing which observations are to be derivable, reintroduces the original problem. For if the observations to be derived are unspecified, the auxiliaries needed for their derivation are indeterminate. The original problem is just that theories do not, in general, have observational consequences, except in conjunction with further theoretical assumptions. The attempt to satisfy both EE and UD with respect to these assumptions, by proposing a condition of conformity with observation to select them, effects no constraint. That requires an independent criterion of epistemic warrant for their eligibility, in violation of UD. It is only because we already have warranted belief in some theories that we can identify any cases of empirical equivalence among theories. Instead of a threat to my case for the possibility of warranting theoretical beliefs, EE appears a confirmation of it.

The problem that the role of theoretical auxiliaries poses for EE might be circumvented by finding a way to establish that theories are empirically equivalent independently of determining their empirical-consequence classes. We would need to be able to ascertain that the classes are the same, without determining what the classes are. However, there would still need to be a fact of the matter—known or not—as to what the classes are, and that requires fixing the auxiliaries. Empirical equivalence established in such a way would still be indexed to the state of evidence that warrants the auxiliaries, and there would be no general guarantee that changes in evidence would sustain it. Thus EE would not be established, at least not in the original form that seemed to entrain UD. One could resort to a version of EE maintaining that the consequence classes are bound to change identically with changes in available auxiliaries, or to a version allowing that while every theory has *some* equivalent rival, *what* rival this is can change. But these versions would seem to be ad hoc, motivated only by an antecedent interest in protecting EE from the limitation of indexing and lacking independent support. There is no particular reason to expect the development of auxiliary knowledge to preserve empirical equivalence in either of these ways.

The only general strategy for upholding EE that is capable of coping with this difficulty is algorithmic. This strategy exists in many versions. Usually, the empirical-consequence class O of an arbitrary theory T is supposed to be given, and the algorithm operates on T to produce another theory T' whose consequence class is also O. The problem is that the T' that the algorithm produces is usually not a rival of T. For example, a T' that existentially generalizes over the theoretical entities purportedly referred to by T, to obtain O while avoiding T's particular ontological commitments, is implied by T. It is therefore acceptable to anyone who accepts T and is no rival. Moreover, it continues to carry theoretical commitments, albeit evasive ones. Nor is a T' that conjoins metaphysical baggage without empirical consequence onto T, and so implies T, a rival. There is no question, in such cases, of deciding *between* T's and T''s explanations of O, T''s explanation, if any, being parasitic on T's. But if T' is not a rival, EE fails. What the algorithm delivers is only the admonition that the observations could be just what they are even if T were false, that they are strictly compatible with the falsity of T. And this much is trivial. Realism does not maintain that theories are deducible from observations.[19]

Perhaps the proposition that T is false, and yet everything observable happens as if T were true, is a rival. At least this proposition is logically incompatible with T. But if it is a rival, this surrealist transform of T is surely not a theory. It is in danger of collapsing into O, as discussed in chapter 1. Theories, on my propositional view of them, must provide some explanatory analysis of an independently identifiable body of attestable facts. It is not *theorizing* about such facts merely to generalize or compile them, while stipulating that a particular theory of them be false. An EE salvaged by the observation that as O entails itself, it is itself an empirical equivalent to T, is one in which I readily acquiesce. Its epistemic import is nil.

Of course, one could add to one's stipulation that T is false, while everything observable happens as if T were true, the "theory" that some (very resourceful) agent deliberately makes all this the case. The world systematically unfolds as if T were true by some means that T wholly misrepresents. What this proposal adds to the claim that T just happens to agree, in its observable consequences, with the true theory—a claim which I have argued is not explanatory—is a connection between T and the purportedly true theory that does not involve similarity or representativeness. We observe what we do *because of the content of* T—not because that content accurately represents what is producing the observations, but because the condition that T get the observations right operates as a constraint on the actual mechanism of production. This is Descartes's "evil-demon" argument, which can be advanced against any ampliative inference from experience. Its explanatory resources are exhausted by its attribution of intentionality to the posited agency. These are comprehensible only by (distant) analogy to the case of human agency, and human agency is explanatory only to the extent that we understand the biology and physiology of human beings. We need some theory of how Descartes's demon does it, a theory that does not itself admit of observable consequences other than those obtained from extant physical theories, if we are to take this proposal seriously as a competing theory. More important, this proposal's threat to undermine theoretical beliefs is no greater than its threat to undermine ordinary perceptual beliefs, whose vindication is not my burden.

On the other hand, one might not stipulate that T be false, but offer reasons to think T false that are not at the same time reasons to dispute T's empirical consequences. André Kukla mentions that the physicalist rejects intentional psy-

---

19. Note that the T''s that algorithms deliver are not distinguishable from T on the semantic view of theories considered in chapter 3. That view identifies theories with classes of models, and applying the algorithms to T does not affect its models. In accordance with the argument of chapter 3, I find this result to be so much the worse for the semantic view of theories. Van Fraassen, whose antirealism depends on EE, is limited to generalizing it from cases, as his commitment to the semantic view prevents him from invoking algorithms.

chology but not its empirical consequences.[20] This would be a curious example with which to champion the algorithmic approach to EE, for the physicalist has no rival theory to offer that reproduces the empirical consequences of intensional psychology. The mere conviction (hope?) that such a theory is possible does not support a version of EE that promotes UD. More important, the supposition that T is rejected for reasons rather than by stipulation implies that the choice between T and the alternative T' thereby generated is not evidentially underdetermined, even if T and T' are empirically equivalent. The example is a reminder that UD does not follow from EE, because theory evaluation may be informed by evidence that does not come from the empirical-consequence classes of the theories being evaluated.[21]

The same objection applies to another algorithm that Kukla thinks can withstand my criticisms.[22] Let T' be the proposition that everything happens in accordance with T whenever observations are being made, and not otherwise.[23] As there are *possible* observations with respect to which T and T' diverge, this algorithm cannot be used to establish EE; its threat is to establish UD directly. But it does so only if evidence relevant to the choice between T and T' is limited to T's observational consequences, whereas there are plenty of strong reasons, of a less direct kind, to deny that the world is subject to unknowable and random suspensions of natural law. What is worse, the very coherence of this T' is dubious. Like all the other algorithms, this one defers to T for a determination of what the allegedly common class of evidential consequences is to be. But the theoretical auxiliaries thereby invoked are general laws whose applicability is not indexed to times of testing. One cannot both assume those laws to determine what observational results T' predicts, and also deny them as part of the content of T' itself. T' posits wholesale departures from natural law coincident with the cessation of observation, followed by a wholesale reversion to natural law, together with the reinstitution of conditions that would have prevailed had the laws never been disrupted, coincident with the onset of observation. And these shifts must themselves be inexplicable in principle, if the algorithm is to be universally applicable. For it could not be applied to a theory that explains them. Certainly, the epistemic constraints on the auxiliaries available to T do not admit such suppositions.

---

20. André Kukla, "Does Every Theory Have Empirically Equivalent Rivals?", *Erkenntnis*, 44 (1996):137–166.

21. See "Empirical Equivalence and Underdetermination." Although I do not, in arguing for MER, elect to exploit such disparate sources of evidential support, I acknowledged them in chapter 5 and am happy to deploy them against the inference from EE to UD.

22. This algorithm descends from one that Kukla proposes in "Laudan, Leplin, empirical equivalence and underdetermination," *Analysis*, 53 (1993):1–7. Laudan and I reply to its original form and propose the present version in "Determinism Underdeterred" (in the same journal and issue, pp. 8–16).

23. Of course, this T' is parasitic on T, and so its status as a rival theory is questionable. Kukla wants to eliminate reference to T by directly specifying what the empirical consequences are to be. But the determination of what to specify can only be made by reference to T. This is the point of the charge of parasitism. Whether or not reference to T is made in identifying its purported rival is not the proper test of parasitism.

With all the difficulties that the role of auxiliaries causes for EE, it is inevitable that someone should propose a version of EE denying that role altogether.[24] UD does not undercut the use of EE to defend it if no auxiliaries are required to generate observational consequences. Of course, auxiliaries are required for actual theories, so we are to imagine coming into possession of an ideal "total theory," or "system of the world," that predicts *all* observations, a theory that embeds all actual theories thought to be correct (and then some). Complete *ex hypothesi*, such a theory contains all resources necessary for prediction and is not supplementable by auxiliaries independent of it. The empirical equivalence of two such theories would not be defeated by UD, because the relevant observational consequence class is specifiable simply as the totality of all true observation sentences. Here are five reasons not to take this approach to EE.

1. There is no reason to believe that the universe does admit of even one total theory. The best surrogates for such a theory in actual science rely on theoretical auxiliaries in crucial ways, and are not foreseeably independent of them. "Theories of everything," in which much information that currently has collateral status in prediction and explanation becomes deducible from deep, abstract principles, still have time-symmetric physical laws and require auxiliaries to get temporal direction. Quantum uncertainty prevents even a theory of everything from predicting phase transitions in the evolution of the universe, on which the present observable state of the universe depends. Quantum uncertainty itself functions as an auxiliary in many explanations, such as that of the local inhomogeneity of the universe. A theory of everything that incorporates quantum uncertainty must still enlist, as auxiliaries not implied by the theory itself, assumptions about processes allowed but not required by quantum uncertainty to get the inhomogeneities we observe. All such auxiliaries involve substantial theory and cannot be exempted from UD in any principled way. There is no basis in the impressive growth of the empirical scope of fundamental physics to project "totality" in the requisite sense.

2. The existence of one total theory is no basis for assuming the existence of another. Arguments against recourse to algorithms apply, for the algorithms used to defend EE are indifferent to scope of application. One possibility is to invoke Duhem's and Quine's theses. A "near total" theory, which clashes with but a single observation, must, according to these theses, be completable in different ways. By Duhem's thesis, there must be alternative deletions that reconcile the

24. Carl Hoefer and Alexander Rosenberg, "Empirical Equivalence, Underdetermination, and Systems of the World," *Philosophy of Science*, 61 (1994):592–608. Hoefer and Rosenberg claim a precedent in Quine and van Fraassen for interpreting EE their way. They are right about Quine. But since Quine holds that the option to take advantage of holistic underdetermination to generate empirically indistinguishable rivals is as rational as historical theory changes that are paradigms of progress, Quine is also committed to the form of EE that I have been criticizing. They are certainly wrong about van Fraassen, who uses EE to support instrumentalism with respect to *all* theory. A theory's empirical adequacy, in van Fraassen's sense, does not obviate the role of auxiliaries. It requires not that the theory imply all true observation sentences, but that all the observation sentences it implies be true. This simple misunderstanding is evidently responsible for the attribution to van Fraassen.

theory with the observation. By Quine's thesis, there must then be alternative supplementations that enable the theory to predict the observation. The result would appear to be alternative total theories.

This holistic scenario does not guarantee totality, however. The Quinean supplementations are supposed to have auxiliary status, which is not permissible within a total theory. And the Duhemian deletions will, in general, sacrifice other observational consequences; these would then have to be reobtained, reintroducing the original problem. But even if totality could be reached, there is no reason to expect the resulting total theories to differ appreciably. The holistic scenario undermines the epistemic import of applying EE to total theories, because it leaves the overwhelming body of theoretical knowledge unrivaled, disarming the threat of UD. Nor could it have done otherwise. If novel success supports realism for partial theories, then an underdetermination inferred from the empirical equivalence of holistically generated total theories must be greatly restricted. The novel success of partial theories is grounds for projecting their survival in the intersection of any eventual, empirically equivalent total theories.

3. The reasons for supposing UD to defeat EE for actual (partial) theories count generally against EE with respect to theories *not known* to be total. Unless theoretical completeness is assumed, the need to develop new auxiliaries, and to invoke them in prediction or explanation, is a perennial possibility. But short of knowing that the universe is at an end, we cannot be in a position to affirm the totality of any theory, however comprehensive. Epistemic conclusions drawn from an EE restricted to total theories must remain hypothetical and tentative. If EE becomes a serious issue for epistemology only at the level of total theories, then it never becomes a serious issue.

4. Even if EE with respect to total theories is safe from UD, UD is no more inferable from it than from EE with respect to actual theories. Commonality of evidential support is not, once again, collapsible into commonality of observational consequences. If, for example, the difference between prediction and retroactive assimilation of observations is an evidential difference, the move from partial to total theories, which affects only comprehensiveness, brings us no closer to UD.

5. Objections to EE that are independent of its conflict with UD apply to total theories. For example, changes in observational status need not reflect changes of theory. The range of the observable is frequently extended by technological innovations pursuant to established theory. Even given a total theory, more phenomena could become observable, affording new opportunities for observational discrimination of rivals. Thus the empirical equivalence of rival total theories cannot be guaranteed to be better than temporary.

Faced with these five objections, one might resort to a weakened conception of total science, on which totality is not *finality*; it does not guarantee immunity to empirical refutation. Let a total science *at a time* be the conjunction of a partial theory with all the auxiliaries permissible at that time.[25] Such a science

---

25. This is Kukla's ploy in "Does Every Theory have Empirically Equivalent Rivals?".

may imply observation statements that turn out false, or may fail to imply ones that turn out true, and so give way to a different total science at a future time. But as the observational consequences of such total sciences are determinate, a state of empirical equivalence among them, should it ever obtain, is permanent. The sciences themselves are (collectively) defeasible, but their empirical equivalence is not. It seems that UD cannot undermine EE with respect to temporally indexed total sciences.

The problem, however, is that if UD is correct, there are no *permissible* theoretical auxiliaries to be conjoined with partial theories into total sciences. The available evidence cannot, by UD, warrant the assumption that any such auxiliary is correct. And, once again, we cannot drop the requirement that auxiliaries be epistemically warranted without depriving EE of all epistemic significance. Instead of producing indeterminateness of the empirical-consequence classes that must be compared to establish empirical equivalence, underdetermination now produces indeterminateness of the very theories whose empirical equivalence is supposedly protected against changes in auxiliary information. The later result is, needless to say, as damaging as the former.

The only surviving challenge to MER that I can find in appeals to EE is its reminder that it is always possible for a new theory to reproduce the phenomena. If ruling out all possible rivals were a precondition for warrantedly interpreting a theory realistically, this would be a serious challenge indeed. For some potential rivals may be unidentified. But unless one takes an exclusively eliminativist view of the weight of evidence, restricting it to the disqualification of theories, covering the possible field of rivals is not a *pre*condition for singling out a particular theory as deserving of realist interpretation. According positive weight to evidence, as per my analysis of the probative force of novel results, enables one to obtain warrant for a given theory *in advance* of worries about possible rivals.

Such rivals, should they appear, have the potential to undercut this warrant, as virtually all beliefs are potentially defeasible. In interpreting a theory realistically, we do commit ourselves to the falsity of any such rival, insofar as it conflicts with the explanatory mechanisms responsible for the novel success that warrants the given theory. But this commitment is an *inference* from the given theory, not something that must, *per impossibile*, be warranted independently. Insofar as we have reason to believe a theory, we have reason to disbelieve that equally good rivals will be forthcoming. And that reason novel success provides. The truth of a theory implies the falsity of rivals yet unknown. On a positive theory of warrant, there is no difficulty in assigning truth values to propositions not entertained. We establish truths not by falsifying alternatives, but by warranting beliefs. Truth, falsity, and all their entanglements, are incidental to realist reasoning. We gather evidence, evidence warrants beliefs, and truth enters the picture because to believe is to believe to be true. The focus is warrant, not truth. Truth comes out at the end, not in at the beginning.[26]

---

26. This priority my realism shares with the explanationist epistemology of William Lycan in *Judgment and Justification* (Cambridge: Cambridge University Press, 1988).

Is There Truth in Virtue?

Of course, one may believe the empirical consequences of theory T without believing T. That is, one may believe that every element of the set T^* of T's observational consequences is true without believing that T is even partially true.[27] Believing T in addition to (the elements of) T^* increases epistemic risk. Is there any compensation for running this increased risk? The compensation that one would expect for running an epistemic risk is the chance of being proved right, or at least of garnering support. In this case, that compensation does not seem to be available, for the only evidence that bears on T does so indirectly via T^*; it does not seem possible for evidence to bear differentially on T. So although the risk run in believing T^* is compensated, the risk run in believing T *in addition to* T^* is not compensated. That is, the component of belief that represents the difference between believing T and believing merely T^*, $T - T^*$, is uncompensated. If we label this the *theoretical* component, the moral seems to be that nothing is gained by believing theories.

This is the conclusion that van Fraassen champions.[28] He advocates T^* over T as the proper object of credence. More precisely, his "constructive empiricism" claims that the strongest warrantable endorsement of any theory is the endorse-

---

27. I do not suggest that anything should incline us to act on this liberty. I myself experience great difficulty in imagining what would impel one to do so. My problem is that some antirealists purport to find it easy, although they are not exactly forthcoming with well-motivated examples. The case of intentional psychology strikes me as being as likely as any, and in that case I do not find T^* at all credible. Acknowledging the superiority and indispensability of folk psychological concepts for ordinary explanatory purposes is a far cry from presuming that the predictions they issue never err. In holding that physicalism, while far from developing a theory that can vie with intentional psychology in the prediction and explanation of behavior, must nevertheless be correct, I hold that the conceptual discriminations of intentional psychology are too rough and loose to be fully dependable in complex situations. I have a much better grip on T in this example than I do on T^*.

My imagination does a bit better in the abstract. Suppose we have competing theories that are both empirically equivalent and empirically adequate *to the best of our knowledge*. Suppose that knowledge is considerable—a great many observable consequences of both theories have been confirmed; no known observable consequence of either has failed; and no consequence of either has been identified that is not a consequence of the other. Then we might decide that there is (mounting) evidence of the empirical adequacy of each that cannot warrant belief in the truth of either. This scenario is certainly a possibility, but it is subject to two reservations: The difficulties found to affect EE suggest that it is too unlikely a scenario to generalize. And the inferences to the empirical adequacy and empirical equivalence of the theories, from the truth of observable consequences identified and tested to date, violate limitations on enumerative induction. In chapter 5, I rejected induction in the absence of an explanatory connection, which T would here have to provide. A reason not to believe T is a reason not to accept the observations accumulated to date as an inductive basis for projecting empirical adequacy.

Suppose, however, that we do believe one of the theories, and that this theory implies that the other is both false and empirically adequate. I agree that this is a case in which T^* is believable while T is not. But as this case depends on theoretical belief, it cannot be used to elevate the credibility of T^* over T, in general. Moreover, in any real case, any reason we would have for believing one of the theories would probably also be a reason not to believe that the theories are empirically equivalent.

28. See *The Scientific Image*, and van Fraassen's reply to critics in Paul M. Churchland and C. A. Hooker, eds., *Images of Science* (Chicago: University of Chicago Press, 1985).

ment of its empirical adequacy, the truth of its empirical consequences. It is not simply T^* as such that, in the best case, one believes, for the contents of T^* may be identified only by reference to T; rather what one believes is $A(T)$, the thesis that T is empirically adequate. It is to be emphasized that this is a belief *about* T, not simply a purported recognition of certain observational facts. If it were only observational facts that were believed, the question of why the belief is true might be answered by something short of commitment to T. The claim that certain observational facts are consequences of T, and that T is reliable as an indicator of the observable, might be thought to answer it.[29] But the question of why T is empirically adequate cannot be answered in this way; T's reliability just *is* its empirical adequacy. The only way to explain T's reliability is to impute some truth to $T - T^*$, to suppose that the theoretical mechanisms actually responsible for the truth of T^* are represented, to some degree of accuracy, by $T - T^*$. As that supposition is an uncompensated risk, constructive empiricism demands that T's reliability be left unexplained.

The obvious premise to challenge in this reasoning is the claim that evidence for T^* exhausts the possible evidence for T.[30] The warrant I expect for T comes not from the elements of T^* as such; indeed, I hold that in general the elements of T^* *do not* warrant T. Only for *novel* elements do I claim epistemic warrant. And then it is not the results as such, but *the fact of their novelty* that gives them evidential status. This fact depends on their relations to other theories as well as to T, and on facts about T's provenance. My view therefore violates van Fraassen's constraint on evidence.

The question, then, is whether novel status can be evidential. In arguing that it is, I appealed to the truth of a theory as the only possible explanation of its novel success. I thereby treated explanatory power as an indicator of truth.[31] Ex-

---

29. This answer was faulted in chapter 1, however. Recall the discussion there of the explanatory resources of the hypothesis of empirical adequacy.

30. This claim may be faulted on the same grounds that the inference from EE to UD was originally faulted: The evidence relevant to assessing a theory is not coextensive with the theory's observational consequences. It will increase the scope of my argument not to take advantage of this thesis here. Indeed, it would be inappropriate for me to do so, as my own defense of realism is independent of it.

31. Note that this is not the same as treating explanatory power itself as evidence, additional to that which the novel results explained already provide. It is not as though one first tallies up the evidential weight of novel results, and then adds to this weight in recognition of T's ability to explain those results. Van Fraassen makes much of the incoherence to which such a reckoning leads in the context of a probabilistic epistemology (*Laws and Symmetry* [Oxford: Clarendon Press, 1989], pp. 160–169). Rather, novel status is reflected in the *initial* allocation of evidential weight; novel success counts more than mere conformity of observation to theory, and this is so for explanatory reasons. As shown in chapter 5, explanatory considerations must affect the evidential weighting of observations for purposes of ampliative inference.

Van Fraassen's demonstration of the incoherence of inference to the best explanation cannot be applied to my argument, because my conditions for novelty preclude the situation he assumes, in which competing hypotheses explain the same evidence. More generally, because I want the epistemic evaluation of theories to wait on the determination of their performance as novel predictors, I do not follow him in assimilating "better" for explanations to "more probable." This is the *only* respect in

planatory power is an example of what van Fraassen calls "theoretical virtues"; others are simplicity and unifying power. His thesis about them is that their importance is entirely pragmatic; they are not epistemic, not indicators of truth. One reason he gives for this thesis is that a belief's possession of such virtues does not increase its epistemic risk. As nothing is gained if nothing is ventured, they offer no compensation. I did not simply assume that explanatory power is evidential, however. I defended the epistemic importance of novelty by arguing that ampliative inference in general has an abductive component. If this is correct, then having denied that explanatory power can be evidential, van Fraassen is not entitled to make even T^* an object of credence, let alone $T - T^*$. For T^* represents an ampliative extrapolation beyond any possible body of available observational evidence. The alternative to realism becomes not constructive empiricism but Humean skepticism.

Let me delineate this line of reasoning more finely. I cannot come to believe T^* simply by induction from observations. I must use T to determine what beliefs are constitutive of T^*, what believing T^* commits me to. I might use T for this purpose, not because I regard T itself as an object of belief, but because I regard $A(T)$ as an object of belief. That is, I might regard the condition of needing to be true if T is true as an indicator of potential epistemic commitments, not because T is a potential epistemic commitment (even though it meets this condition), but because its empirical adequacy is a potential epistemic commitment. $A(T)$ directs me to the same empirical phenomena that T does.

But how can I come to believe $A(T)$ without inferring it from T? The answer would have to be that having used T to identify what beliefs the belief that $A(T)$ consists in, I then induce those beliefs directly from observations without further involvement of T. But I cannot so dismiss T at this stage, for I require its explanatory power to perform the inductions. I cannot defer to the weaker claim $A(T)$ for the needed explanatory power, for that claim simply restates what is to be induced, adding that it follows from T. Even if $A(T)$ is taken to explain the individual observational facts that I induce from (which, in chapter 1, I would not allow), it cannot be taken to provide the explanations I need to perform the inductions from them, for it itself is the conclusion of these inductions. The empirical adequacy of a theory that requires emeralds to be green does not explain why emeralds are green; only the theory itself explains this.

By contrast, T can at once be the conclusion of an ampliative inference from its novel results and explain these results, because the inference to T is abductive. It proceeds not from the results, as such, but from their novelty; even if the results, as such, can be explained without imputing truth to T, because of their novelty T's ability to predict them successfully cannot be so explained. $A(T)$ cannot explain why T is successfully extendable into new domains, as novel success

---

which the hypotheses of his examples are better explanations. I submit that if a hypothesis is better only in this respect, then it is absurd to represent its explanatory superiority as an elevation in its probability beyond what the evidence determines, and it is obvious that doing so violates the Dutch Book standard of coherence.

implies; it simply *is* the assertion that T will be empirically successful wherever applied. Even if I explain what has been observed so far by appeal to $A(T)$, I do not explain the generalizations to be induced from what has been observed so far by appeal to this hypothesis, for these generalizations, collectively, *are* this hypothesis (or a proper part of it). I can no more come to believe $A(T)$ just by induction from T's adequacy to date, without explanation of this adequacy, than I can come to believe the generalizations comprising T^* just from their instances, without explanation. Only $T - T^*$ provides the explanation that induction to $A(T)$ requires.

So what is to be made of the claim that believing $T - T^*$ is an uncompensated risk? The first thing I make of it is, obviously, that if the claim is both true and sufficient to defeat belief in $T - T^*$, then belief in $A(T)$ is also defeated. For we lack the machinery to warrant belief in $A(T)$ if uncompensated explanatory beliefs are disallowed and $T - T^*$ is such a belief.

Second, I deny that the claim is sufficient to defeat belief in $T - T^*$. $T - T^*$ is an artifact of the division between observation and theory. It is frequently supposed that it is the realist who needs this division to formulate his thesis.[32] The present discussion shows that the antirealist needs it to circumscribe the range of his skepticism, to protect himself from too sweeping a skepticism. The realist wants to provide warrant for T as such, unrestricted as to component commitments. He acquiesces in $T - T^*$ only because, given its formulation, T commits him to it. T is the realist's proper object of warrant, and belief in T has compensations. That belief is vulnerable to refutation and amenable to support.

Do we really want to endorse an epistemological principle that proscribes believing any proposition subdividable into components not all of which are individually compensated? Such a principle proscribes believing T^* as readily as T. For example, T^* subdivides into T^*_1: All of T's observational consequences hold whenever observations are being made; and T^*_2: All of T's observational consequences hold whenever it is not the case that observations are being made. What compensates believing not only T^*_1 but also T^*? What is to be gained by incurring the extra epistemic risk of believing T^*_2? T^* is no more vulnerable to refutation by observation than T^*_1, and is therefore no more amenable to support. Yet van Fraassen, who thinks that observations alone have evidential relevance, that theoretical virtues like explanatory power—not to mention novelty—are entirely pragmatic and have nothing to do with warranting belief, does not wish to let T^*_1 delimit the scope of warranted belief. There are any number of directions of possible retreat to lower levels of epistemic risk that he will not follow, and with good reason. It is obviously inappropriate to demand that belief in any proposition deducible from believed propositions be individually warranted. What is the individual warrant for believing the proposition, deducible from T^* and therefore believable by van Fraassen, that those of T's observational consequences that will never be tested are true?

The trouble with van Fraassen's epistemological principle is that it ignores the contexts in which beliefs are formed. Whether a belief requires its own distinctive

32. Laudan makes this assumption in *Science and Relativism*.

compensation depends on what background beliefs, already compensated, it issues from and how it issues from them. Ampliative extensions of warranted beliefs should offer further compensation. If I first warranted T^* and then extended my credence to $T - T^*$ by ampliative inference, the demand for further compensation would be appropriate. But this is not how theoretical beliefs are formed, nor how (my) realism would have them formed, nor could it be; it is a reconstruction of their formation that suppresses the actual course of reasoning. $T - T^*$, insofar as it is an isolable object of belief at all, is inferred not from T^* but from T^*'s inductive basis, the empirical reliability of T to date (with respect, I would have it, to novel phenomena). In this capacity, $T - T^*$ certainly carries its own compensations. It is from $T - T^*$ that T^* is then inferred, but nonampliatively and so without the requirement of additional compensation. Because of the fundamentality of abduction, this is the only viable route to T^*.

My third rejoinder to the claim that believing $T - T^*$ is an uncompensated risk is to deny that believing $T - T^*$ offers no compensations beyond those of T^*. $T - T^*$ does not carry observational commitments beyond those T^* carries, and so, given its deductive relation to T^*,[33] can be no better supported by observations than T^*. Nevertheless, it is vulnerable in ways that T^* is not. For example, if I believe not only T^* but also $T - T^*$, then I do not foresee the development of successful, rival theories to T that provide as good an explanation of T^* as T does. I expect, minimally, that any such rival will fail to achieve a comparable record of novel success, and, maximally, that no such rival will be forthcoming at all. I should be willing to predict that any such rival will face empirical failures. I should maintain that an apparent empirical equivalence of such a rival with T will prove temporary, that possibilities for their empirical discrimination, to the rival's detriment, will arise.

What justifies such expectations is just what justifies a realist interpretation of T to begin with: the novelty of T's successes. I can be proved wrong; events that do not threaten T^* could compel me to admit error. My compensation, therefore, is the chance that events will bear me out, that $T - T^*$ will survive sustained inquiry as the only viable explanation of T^*, while compiling an ever-more-impressive record of novel success. The constructive empiricist runs no such risks and is eligible for no such rewards.

Of course, it is not necessary for the constructive empiricist to deny that endorsing theory increases epistemic risk. His point could be (and van Fraassen's originally was)[34] that while the risk is epistemic, the compensation is purely pragmatic, or that the risk outruns the compensation and so is irrationally incurred.

---

33. Remember, from the discussion of empirical equivalence, that the first condition alone is insufficient.

34. Van Fraassen's antirealism shifts considerably from his book *The Scientific Image* to his essay in *Images of Science* to his book *Laws and Symmetry*, attenuating in the process. The latest position restricts issues of rationality to changes of belief; a question of rationality is properly raised only when an already-established belief system is to be modified. This view affords no general indictment of realist beliefs as irrational, even if nothing supports them over corresponding claims of empirical adequacy.

But again, short of a sweeping skepticism, there can be no general epistemic principle proscribing increases of epistemic risk. Far from licensing A(*T*), such a principle would impugn ordinary perceptual beliefs—at least those not restricted to occurrent experience.

Nor is it true, in general, that explanatoriness increases risk. This only seemed true in the context of comparing *T* with A(*T*), and A(*T*) has been found not to be the proper basis of comparison. *Among theories*, explanatoriness need not increase risk. A theory that adduces a single unifying principle to explain a number of seemingly disparate laws is more explanatory, but no more epistemically risky, than a theory that propounds a number of independent hypotheses to the same effect. In the context of this comparison, the claim that the greater explanatory power of the unifying theory is a virtue indicative of truth cannot be faulted by a general proscription of risk. But this is just the comparison that the realist finds pertinent. He supposes that *something* explains the empirical phenomena that science is wont to theorize about, and, seeking grounds for credence in some such thing, he wants explanatory power to count. Naturally, he knows that, strictly speaking, it is unnecessary to theorize at all; that is not at issue.

The realist can go further. Whether or not it is necessary to theorize at all depends on one's ends. Perhaps the strict empiricist need not theorize at all, but is theorizing incidental or optional to the epistemic ends *of science?* Van Fraassen thinks so. He contends that theories serve only pragmatic ends, and that belief in A(*T*) is all the belief that scientific method requires. No doubt, there are features of scientific practice that proceed quite well amid a systematic suspension of theoretical commitment. But to make a case for the general independence of method from theoretical commitment, one would have to examine a rich range of practices that have been found to advance scientific ends and show that none of them owe their rationale or execution to any theoretical belief—not only not to the truth of such beliefs but also not to their possession. Whether van Fraassen has done enough of this to make a credible case is questionable.[35] For example, I have argued that there are both systematic features of scientific method and specific programs of successful research that make no sense unless beliefs in the reality of theoretical entities, and in the possibility of determining their properties empirically, are attributed to researchers.[36] Here I will add to that argument but a single example of an inability of constructive empiricism to rationalize scientific practice. It is, however, an example from which it is difficult not to generalize.

The "standard model" of elementary particles has recently been completed by the discovery of the top quark. Since the early 1970s, there has been abundant evidence from electron scattering experiments that positive electrical charge is not continuously distributed within protons but is highly localized. Much in the

---

35. Perhaps the most controversial part of van Fraassen's argument is his account of the search for unification. It is not clear that the insistence on finding unified theoretical accounts of phenomena for which existing theories are already adequate, so far as we know, is explainable on constructive empiricist grounds.

36. Jarrett Leplin, "Methodological Realism and Scientific Rationality," *Philosophy of Science*, 53 (1986):31–51.

way that Ernest Rutherford was able to show that positive charge is localized within the atom, these experiments argue for a corpuscular internal structure to hadrons—that is, for the existence of quarks. It is fair to say that most particle physicists were confident of the existence of the remaining top quark, although, in their ignorance of the energy required to reveal it, they were divided in their expectations for its detection. But this is a recent attitude. Quarks were introduced for purely theoretical reasons, prior to any experiments interpretable as their detection. In the context of their first introduction, by Murray Gell-Mann and George Zweig in 1963, their purported ontological standing was unclear. This initial unclarity was not so much a question of their existence, as it was a question of whether it was appropriate to treat them as existential posits at all. The former question is naturally unresolved so long as the relevant theory lacks sufficient empirical corroboration (and is forever unresolved if, as antirealisms maintain, no experimental corroboration is ever sufficient to warrant any theory). That question may as readily be addressed to hadrons, as to the quarks that compose them. The latter question pertains to how the relevant theory should be interpreted, irrespective of its evidential support, and in the present example it is specific to quarks.

Quarks were originally a mathematical scheme for ordering and classifying the vast variety of hadrons, the particles subject to the strong force. They need not have even been *intended* as physical posits, let alone have turned out to be detectable. One could have used quark theory as a formal classificatory scheme to obtain conceptual leverage on the complexity of types and properties of hadrons, without attributing, *even in theory*, an internal structure to hadrons. This is how constructive empiricism would have us understand the function of *all* theories: Theories are not proper objects of credence; their virtues are purely pragmatic. The trouble is that here the pragmatic utility of the formal scheme is directed not at observable entities, but at entities that are themselves theoretical. If we treat all theories as pragmatic devices only, as incapable of serving even as candidates for realist interpretation, whatever the evidence, then how do we make sense of the difference between our two questions? How do we reproduce the distinction *made within science* between theoretical, physical posits, like hadrons, and conceptual devices not, or not necessarily, intended as physical posits, like quarks, which originally could be construed as having only formal significance? The issue here is not one of warrant for theoretical beliefs, but of the role of theoretical beliefs in the practice of theorizing. If, as constructive empiricism maintains, there is no such role, how are we to understand what theorists are doing when they debate the proper status of quarks, prior to considerations of empirical warrant?[37]

37. Not only antirealisms, but any views that refuse to allow theory to be a proper object of belief, have difficulty with this sort of question. For example, Ian Hacking's "entity realism," developed in *Representing and Intervening*, makes a realist view of theoretical entities dependent not on belief in theory, but on experimental access and technological control. It licenses realism about those theoretical entities that we are able to use to investigate still more theoretical parts of nature. But the distinc-

This debate was serious and prolonged. It had nothing to do with philosophical reservations about the interpretation of electron scattering experiments as the detection of quarks. An adequate philosophy of science must make sense of it. But it does not make sense to represent researchers as puzzling over whether or not entities that they do not take to be real should be regarded as having real or merely fictitious constituents. If hadrons were not to be regarded as real, how could their constituents even be candidates for reality? Constructive empiricism seems to require us to regard the debate as a massive confusion, which a proper philosophy could have dispelled without attention to the details of hadron theory. But a philosophical position that—for strictly philosophical reasons, not scientific ones—represents substantive debates within science as confused is not giving us an adequate account of scientific method. It is not capturing the rationale of theoretical deliberation.

The failing looks even worse when we consider the experimental side of the issue. The experimenters were attempting to determine the scattering pattern of electrons bombarding protons. They wanted to know whether the proton acted as an indivisible unit in deflecting electrons with its positive charge, or whether the deflecting force of the proton was localized within it. The question to be answered was whether the proton possesses internal structure. Just to pose this question requires stipulating the existence of the proton itself. The question of the existence of distinguishable constituents of the proton makes sense only against a background of belief in the existence of protons. How is the experiment to be interpreted, how is its rationale to be reconstructed, if, with constructive empiricism, the existence of the proton is not to be regarded as a proper object of credence? Researchers cannot sensibly be represented as attempting to determine whether or not entities that they do not believe in have constituents that they also do not believe in. Contrary to constructive empiricism, researchers must be credited with theoretical beliefs as a condition of understanding their practice. Much scientific work does not make sense if realist attitudes within science are disallowed.

---

tion between what can and cannot, in principle, be used to press inquiry further is itself made for theoretical reasons. If we cannot believe these reasons, we cannot know what will be a candidate for realist interpretation.

In the present case, the theory of the strong force implies quark confinement, which precludes the sort of experimental access that Hacking thinks we have achieved with respect to electrons, about which he is realist. If we believe this theory, then by Hacking's standard, we must not believe that quarks are real. But if we believe this theory, then we *do* believe that quarks are real. It seems that we are to regard quarks only as conceptual devices, not open to realist belief, but the reason that we are so to categorize them is something that we are also not to believe. I do not know how Hacking can make sense of this. Unless he makes sense of it, he cannot recover the distinction drawn within science between real entity and conceptual device.

The general problem seems to be that the difference between potential experimental accessibility and inaccessibility is contingent on physical features of the world. There is no reason to expect it to correlate with the difference between status as conceptual device and status as candidate for realist interpretation. Without such a connection, experimental accessibility cannot be the criterion of the real.

This example of the inability of constructive empiricism to rationalize scientific practice defends epistemic realism in general, by showing that not just $A(T)$ but T itself must be an appropriate object of credence. To the example I would add a complaint on behalf of my own realist position. The novelty of a prediction makes no difference to its bearing on a theory's empirical adequacy. The correctness of a novel prediction supports a judgment of empirical adequacy no more than the correctness of any observational consequence. That a theory is empirically adequate does not imply that any of its predictions are novel, and does not explain how the theory achieves novel success if any of them are. But empirical adequacy is the strongest epistemic status that constructive empiricism can warrant. Therefore, constructive empiricism can make no sense of the scientific practice of according special epistemic significance to novel results. According to the argument of chapter 5, this failing creates an argumentative burden for van Fraassen's position. Having recommended constructive empiricism for the adequacy of its resources to the methods and aims of science, van Fraassen is in fact in the position of having to fault scientific practice. Neither the prospect of doing this successfully, nor the recovery of his original defense of constructive empiricism if he did it, are open to him.

Let me engage constructive empiricism on behalf of my own position in broader terms. In order to maximize the scope of my criticisms, I have, so far as possible, allowed van Fraassen his way of formulating the issue of realism in its greatest generality. I wish now to recall my differences with this formulation. Van Fraassen thinks that realism treats explanatoriness as a truth indicator. He argues that it cannot be, because a feature that indicates truth must increase probability, whereas explanatoriness decreases probability by increasing epistemic risk. He further argues that explanation is context dependent and, therefore, pragmatic rather than epistemic. The revealed failings of these arguments aside, it is to be emphasized that the case for realism need not be formulated van Fraassen's way. My own defense of realism proceeds quite differently.

Van Fraassen wants to identify the explanatory relation between theory and empirical results (plus context, in his view) first, independently of questions of truth, and then to ask whether the presence of this relation grounds inference to theory. Of course his answer will be "no," because the further step of introducing truth requires further justification, beyond explanatory considerations, which he cannot find. One can stop short of realism, and yet retain the explanatory advantages of theory. But in the case I make for realism, truth itself is part of the explanation. Only the attribution of truth to theories explains their novel success. Without introducing truth, one has explanations (or, as I would prefer, purported explanations) of empirical results as such, but not of a theory's novel success. This is why my explanationist defense of realism does not rely on appeal to the implausibility of invoking the miraculous to explain scientific success. I am not imposing truth as explanation on an independently established, explanatory relationship between theory and empirical success, but invoking it to explain certain empirical successes directly.

Van Fraassen's formulation invites the antirealist slogan that in the absence of independent access to the truth, we have no basis for estimating the likelihood that false theories will be successful.[38] If it cannot be pronounced unlikely that false theories will succeed, then its success cannot impel us to credit a theory with truth. If we observe instances of the relation of theory to empirical success with no criterion for attributing truth but empirical success itself, then we cannot compare the frequencies with which true and false theories enter this relation. I can reply that this point does nothing more than invoke, once again, the possibility that success occurs by chance, which is not in dispute. The legitimacy and importance of seeking explanations are not contingent on prior demonstration that the phenomena to be explained could not be accidental. But I do not rely on this reply. Instead, I undercut the whole line of argument by reconceiving what is to be explained. It is the fact that a theory is a successful, novel predictor that wants explanation, and the theory itself does not explain this. The theory itself explains only the empirical phenomena, not why *it* should be successful at predicting them. We explain this by attributing truth to the theory, in case the phenomena are novel for it; we do not find the explanation of this in the theory.

The alternative to my realism is that one has no explanation of novel success at all, whatever one's explanatory resources with respect to much else in science. That is, once again, a possibility; one may resist realism by holding that certain aspects of science are simply inexplicable. But in my approach, truth is not simply added on after all the explanations one could reasonably want are in. Truth is part of any explanatorily adequate account of what science does, or, at least, can, achieve. The alternative to invoking it is acquiescence in (a certain limited degree of) mystery. Van Fraassen wants the issue to be whether explanation grounds inference, while I want it to be whether explanation is to be sought in the first place.

NOA's Bark Is as Terse as It's Trite

Of course, Fine, in posing his dilemma for realism, is aware that Hilbert's program of finite constructivism in mathematics failed. The inferential methods it allowed are incapable of establishing the consistency even of arithmetic, according to Kurt Gödel's theorem of 1931. But this failure does not deter Fine. He insists that Hilbert's admonition to eliminate from metatheorizing any mode of inference used in theorizing is "correct even though it proved to be unworkable."[39] Fine's retrenchment is, on its face, strange. Surely, finite constructivism is not a correct theory of mathematical proof. Gödel's result brought an end not to proof theory, but to finite constructivism; Gerhard Gentzen's proof of the consistency of arithmetic uses infinitary methods, and proofs of consistency and completeness theorems standardly deploy techniques that are, if anything, less stringent than the inferential methods allowed in the proofs they underwrite. For example, it is standard procedure to use the principle of mathematical induction,

---

38. It also invites a charge of circularity, but this I leave to the next section, where it arises more powerfully.

39. Fine, "The Natural Ontological Attitude," in Leplin, *Scientific Realism*, p. 85.

which is one of those assumptions that agitate protectors of mathematical rigor. On what basis does Fine judge the moral stronger than the argument that sets it?

Played out, the mathematical analogy cautions us not to assume, as Fine does, that an abductive defense of abductive methods is insignificant because trivially guaranteed. There is no general assurance that a standard of warrant will be successfully self-applicable. Popper's requirement of falsifiability, to take an obvious case, is not itself falsifiable. Self-application can be a severe test of a theory of inferential warrant; passing it, though logically inconclusive, can carry significance. It certainly does in mathematics, where nonconstructive consistency proofs are typically far from trivial. Gentzen's proof is both sophisticated and significant, but could be neither if Fine were right. Why does Fine think that the mathematical analogy supports importing constructivism into philosophy?

It is not as though the application of the constructivist ideal to science avoids the difficulties that defeated it in mathematics. One cannot eliminate from the philosophy of science the forms of reasoning used within science, for the simple reason that one would be left without any form of reasoning at all. Neither Fine nor anyone else has identified a single, rationally cogent form of reasoning without exemplification in science. And it is unavailing to limit the constructivist restriction to suspect forms of inference. Every form of reasoning used in science sometimes issues in false conclusions. What will a discriminating standard of suspicion be? We cannot very well restrict philosophy to valid deduction, while at the same time charging it with underwriting the ampliative argumentation that abounds in science. No mathematical moral will make sense of that. The only moral I can extract from this sorry flirtation with logical foundationalism is that there are some a priori presuppositions to rationality itself. The alternative is obvious circularity.

I argued in chapter 5 that straight induction is no more fundamental than abduction, and I must here reject as misconceived any demand that the latter be legitimatized by reduction to the former. Of course, Fine is free to extend his moralizing by dismissing them both from philosophical reasoning about science (although he must then withdraw his endorsement of the skeptical historical induction, which gets his dilemma off the ground). The result is *no* philosophy of science at all—NOA, in effect. Unenamored of minimalism in art or literature, I am also unable to see its charms in philosophy. A minimalist compromise can always be forged among competing doctrines that are not logically exclusive, but the cost is simply abandoning the issues and questions that create philosophy in the first place. NOA is not an alternative philosophy to realism and antirealisms, but a preemption of philosophy altogether, at least at the metalevel. I see no more reason to rule out of court philosophical questions *about* science—about what knowledge and truth are, what justifies beliefs, what is the potential scope of scientific knowledge—than to dismiss substantive questions *within* science. Scientists do not retreat to minimalist positions when faced with competing theories. They try to find grounds for preference because only thereby can their questions be answered. They are not content to proclaim—to borrow an example from Kukla—that the universe expands, eschewing rival big bang and steady-state accounts of how it does this; this "natural cosmological attitude" is not an option for them. Nor is the natural ontological attitude an option for one who seeks knowledge of science.

In particular, the idea that scientific theories "speak for themselves," that one can "read off" of them the answers to all legitimate philosophical questions about science, cannot be squared with the rich tradition of philosophical debate among scientists over the proper interpretation of theories. If the formalism of quantum mechanics "spoke for itself," what were Bohr and Einstein arguing about? Clearly, the formalism does not ordain its own completeness. Why did Minkowski's geometric interpretation of special relativity make such a difference to the theory's acceptability? The answer must be that realism at the level of interpretation was an issue that the theories themselves left unresolved, but whose resolution was crucial to understanding the relevant science. Quantum mechanics, by itself, does not tell us whether the uncertainty relations are a restriction on knowledge or on the world. Special relativity, by itself, does not tell us whether it is enough to relativize space and time or whether space-time must be reified. Fine has made major contributions to such discussions and must realize that, in so doing, he has exceeded the confines NOA would impose.

To turn the tables on Fine's constructivist admonition, why *shouldn't* philosophy use the methods of inquiry and appraisal found most successful within science? This is what epistemological naturalism counsels. And abduction has been eminently successful in advancing theoretical progress. The methodological legitimacy of positing unobserved entities, if controversial to this day in philosophy, has not been controversial within science since the rise of the wave theory of light early in the nineteenth century. Without abduction, we would not have any of the science we now value. Certainly, NOA is not content with the asceticism of a nonabductive world view. It would have us go on talking in a realist vein about scientific conclusions, employing normal referential semantics in giving scientific explanations, treating scientific hypotheses on a par with descriptions of ordinary perceptual objects. Only it would have us beg off when asked what all this commits us to. It happily tells us what electrons are, but balks at telling us what knowledge, truth, and justification are. Its response to these questions is to erase the blackboard. This is unacceptable. You cannot have your science without taking philosophical responsibility for it.

If I am dismissive of NOA, I must take seriously Fine's charge that realism is superfluous. Naturalism advises us that a philosophical theory must not make an exception of itself, but must face up to evaluation by the same standards that it would impose on theories generally. Epistemologically speaking, theories are to be valued in the measure that they successfully predict novel results, supposing that their empirical credentials generally are in order. My response to the skeptical induction protects realism from general empirical concerns; theories that have achieved sustained records of novel success, uncompromised by failure, have not undergone wholesale rejection. But what novel success can realism itself claim?

The answer, I think, is simple. A realist interpretation of a theory is the basis for projecting its continuing success when applied in new domains. Insofar as the mechanisms responsible for the results by which the theory has been judged so far have been correctly identified, such further, testable implications as can be extracted from them should be borne out. The prediction that they *will* be borne out qualifies as novel with respect to a realist interpretation of the theory, for no competing philosophical position delivers it.

Certainly, instrumentalism does not deliver it; instrumentalism gives us no better reason to trust the theory's predictions in new domains than the predictions of a different theory, even one that has so far been unsuccessful. Success in one domain has no epistemic connection with success in another, if its achievement is but happenstance; neither has failure. Of course, the hypothesis that the theory is reliable predicts its continuing success. But this hypothesis is not a competitor to realism; it is a consequence of it. It merely restates the prediction itself. The hypothesis that the theory is reliable but false may look more of a contender. But $T^*\&\neg T$ is neither empirically supportable nor explanatory, as a competing theory must be if its predictions are to interfere with novelty with respect to T. Clearly, the claim of falsity can have no empirical support with respect to a reliable theory. The antirealist wants symmetry here; he wants to say that the claim of truth can have no support either, that reliability alone is supportable. I have been at pains to disabuse him of this misconception. T^* may be independently hypothesized, but it cannot be warranted unless T itself its warranted; to warrant T^* alone would require straight induction in the absence of abduction.

The uniqueness condition requires, for a result to be novel for T, that there be no viable theoretical basis, alternative to T, for predicting the result. Realism with respect to T predicts that T will continue to be successful under further applications. $T^*\&\neg T$, though it implies such success, is not an alternative theoretical basis for predicting it. It is not a theory at all; it explains nothing and is not amenable to empirical support. $T^*\&\neg T$ is, of course, strictly consistent with the evidence that supports T. But it is not viable in the more robust sense—which could be assumed in the less abstract contexts provided by examples of specific theories—of being borne out by the empirical evidence so far available. Evidence that supports T cannot be said to "bear out" $\neg T$. In applying the standard of novelty to philosophical theories, the requirement that there be no reason, independent of T, to predict a result novel for T is perhaps the more perspicuous formulation of the uniqueness condition (and could, granting my view of straight induction, be made retroactive). Clearly, $T^*\&\neg T$ does not give reasons. However one puts it, the case for the novelty of realism's predictions derives from the conclusion, reached in chapter 5, that realism alone explains a theory's novel successes. If no antirealist position explains novel success, none can rationalize the projection of such success.

Realism with respect to T also predicts that no competing theory as good as T, by the standards used in judging T, will become available. As I have argued, antirealisms make no stand on this point. So it, too, may be offered, against the charge of superfluousness, as a novel prediction of realism. Some philosophers appear to want an intermediate position between constructive empiricism—or, more generally, pragmatism—and realism, one on which evidence can support theories without warranting theoretical beliefs. To the extent that a theory is supported by evidence, as opposed to being merely predictively reliable or pragmatically advantageous, there would be a basis for denying the potential for equally good rival theories. Support is presumably a stronger epistemic relation than adequacy; evidence that supports a theory thereby discredits incompatible theories, whether or not it conflicts with their predictions.

For example, Laudan, though no realist, maintains that preferences among empirically equivalent, rival theories can be evidentially grounded, and not merely left to pragmatic criteria of convenience and utility. One reason is that a result that two theories equally predict may be evidence for only one of them, having been presupposed by the other. Another possibility is that a result neither theory predicts may support one of them indirectly in virtue of its connection with a third theory that predicts the result successfully. In general, the empirical equivalence of theories, should it ever obtain, would not imply that they are equally supported by all relevant evidence.

As explained in this chapter, I am with Laudan this far. My problem is understanding what, short of theoretical belief, the evidence is supporting in such cases. Suppose that some evidence supports T over its empirically equivalent rival T'. If realism is to be avoided, some aspect or part of T, not involving its theoretical posits, must be what the evidence supports. There are two constraints on what this can be: It cannot be merely a pragmatic virtue of T, for then preference for T over T', based on the evidence, would not be an epistemic preference; and it cannot be anything that T shares equally with T' in virtue of their empirical equivalence. The challenge to Laudan's antirealism is to meet these two constraints at once.

Predictive reliability meets the first; it is an epistemic virtue that does not carry realist commitments. But it fails to meet the second; unlike predictive efficiency or economy, which are pragmatic, it cannot attach differentially to empirically equivalent theories. Another possibility is the attribute of being well tested.[40] This satisfies the second constraint. But if what is being tested for, what passing tests attests to, is not predictive reliability, then what is it to be but the correctness of theoretical posits? It will not do simply to say that it is T itself that is being tested, and, upon passing, judged well-tested, and to stop at that. This is akin to the pernicious dogma, by which egalitarian relativists enforce intellectual modesty, that IQ tests measure not intelligence but only the ability to do well on IQ tests. Things are tested *for* something, some attribute or capacity; they are not just tested. And one cannot very well test a theory for the attribute of being well tested.

There may be further options, and it is always possible to withdraw the supposition of empirical equivalence, to fall back on the inadequacy of arguments purporting to establish that theories invariably have empirically equivalent rivals. But as I am unable to find any territory for the antirealist who wants to be more than a pragmatist to occupy, I must rest my case for the differential advantages of realism on comparison with pragmatism.

---

40. Laudan's *Science and Relativism* suggests that the strongest epistemic stance that can be warranted for theories is the judgment that they are good at passing empirical tests, and that theories are to be valued for this attribute above all others.

7

The Future of Realism

Limits to Testability in Fundamental Physics

Realism, even of the modest kind that I hold defensible, is very demanding empirically. Results that ordinarily inform theory choice do not support it; they speak, rather, to pragmatic dimensions of appraisal. If realism requires what I require for it, how realistic is it to expect fundamental physics to warrant belief? If the answer to this question has so far, in this book, appeared mixed and indecisive, it becomes downright pessimistic when contemporary directions of theorizing are considered. For some of the latest physical theories are difficult, if not impossible, to test empirically.

Let us be clear about the nature of the problem. It is not that current theories fail to yield empirical predictions. Nor is it that the predictions they yield fail to distinguish them from other theories already tested. Opportunities to meet my conditions for novelty abound in current physics. The problem is that such predictions do not appear to be testable. We lack the technological means to determine whether they are correct. Nor is this problem reasonably regarded as a temporary limitation, comparable to the epistemic situation that has frequently confronted new theories, so that a "wait-and-see" attitude is appropriate. The new situation is that the very theories whose credibility is at issue themselves ordain their own nonconfirmability. If the latest theories are correct, then we should not expect to be able to confirm them. For they tell us, in conjunction with well-established background information, that the conditions under which the effects they predict occur are not technologically producible.

This situation is unprecedented in epistemology. It is all well and good to insist, as a matter of principle, that belief be proportionate to the evidence, and to dismiss as "metaphysical," or otherwise spurious, untestable propositions. But what are we to say when it is the very propositions whose appraisal is at issue that preclude the acquisition of relevant evidence? If the troublesome propositions were the inventions of philosophy or the doctrines of theology, we might dismiss them as intellectual confusion. But what do we do when they come from the heart of the most intellectually progressive mode of inquiry that we know, a mode of inquiry that it was the point of demanding testability in the first place to privilege? The verdict "unscientific" is not an option.

Though unprecedented, this quandary should occasion no surprise. Indeed, it is something of a confirmation of the progressiveness of our science that we have reached it. There was never any reason to expect theories of unobservable phenomena inevitably to admit of observational test, and it was a presumably transient luxury that they did so. There was never any reason to expect technology to keep pace with theorizing as ever-deeper and more abstract physical principles were investigated. To expect theories to be assessable by observation, simply because this is our only purchase upon them, is but wishful thinking.

Current fundamental theories deal with energies that are not foreseeably producible technologically, and whose natural existence is confined to so early an epoch in the evolution of the universe as to be inaccessible telescopically. Either an inflationary stage of expansion wiped out the evidence of what happens at these energies, or this evidence is concealed within regions of space-time not causally connectable with our own. The conditions that prevailed immediately after the inflationary phase are also unrecoverable, due to the indeterminate character of the process of spontaneous symmetry breaking that is thought to have produced the diversity of the particles and forces of nature as the universe cooled. These are all obstacles to testability, if we are to trust some of the very theories that want testing.

Fundamental theory suggests that at very high energies, the particles that carry the forces of nature lose their separate identities, so that a single unifying force replaces the known forces for which distinct theories have been developed and tested. This program of unification has been tested through the level of the electroweak theory, which successfully predicted the W and Z particles. The unification of the electroweak force with the strong force responsible for the stability of the atomic nucleus implies a breakdown in the distinction between quarks, of which nuclear particles are composed, and leptons, which interact by the electroweak force. One consequence of this "grand unification" (GUT) is that the law of conservation of baryon number, for example, which guaranteed the stability of the proton, cannot be fundamental. More precisely, the conservation laws for quark number and lepton number must be approximate only. The transmutation of a quark into a positron, by the exchange of a particle carrying the grand unified force (an "X" particle) with another quark, implies the decomposition of the proton of which the quarks were components. In short, protons decay.

This effect is, in principle, detectable, and at one time its detection was expected. That is, physicists were confident enough of the unifiability of forces to

be convinced, in advance of the anticipated evidence, that the distinction between stable and unstable matter is only a matter of degree. The unexpected
failure to detect proton decay did not weaken this confidence. Instead, advantage
was taken of the option to adjust the unifying theory, extending the life-span of
the proton. Rather than a test of unification, proton-decay experiments became a
measure of the (lower limit of the average) life-span of protons.

More precisely, what happens is that mathematical symmetries maintained by
the separate forces are made to instantiate a common, more abstract symmetry,
and a single, unifying force, with its own carrier particles, is postulated to maintain that symmetry. Failure to detect proton decay is sufficient to disqualify one
particular such comprehensive symmetry (the simplest GUT symmetry, SU(5), as
noted in chapter 5), but other candidates, with the same unifying effect, are available to fall back on. The theorist is committed to the program of unification, not
to any particular implementation of it via a given symmetry. This permits the
reinterpretation of apparently negative findings as guides to the selection of symmetries, with the result that the basic theoretical commitments—to the instability
of matter, for example—lose their vulnerability.

A similar thing happens in connection with another basic theoretical commitment—to the existence of the graviton, the quantum of gravitational force. Arguably, we can detect gravity waves. The orbital decay of pulsars is attributed to their
emission. But the natural occurrence of gravity waves is a consequence of general
relativity that is independent of the unification of gravity with the other forces. If
we are going to quantize gravity and bring it into a form amenable to assimilation
under a larger, unifying symmetry, we must have not only waves but also gravitons, whose spin properties can explain why gravity is always attractive and so,
despite its weakness, can accumulate to govern the large-scale structure of the
universe. The trouble is that gravity is so weak that although there are possibilities
for the indirect detection of gravitons, we cannot interpret the failure to detect
them as disconfirmation of the unifiability of gravity with other, already quantized
forces. As in the case of proton decay, we have opportunities for positive evidential
support, but the underlying theory is not at risk in failing to obtain it.

Similarly, detection of the magnetic monopole would confirm GUT, which
treats electricity and magnetism more symmetrically than did Maxwell's original
unification of them. There ought to be individual units of magnetic charge, as
there are electrons. But the mass of such magnetons is theorized to be so great
that the energy required to make them is unachievable. And the inflationary stage
of the evolution of the universe so reduced their density that the detection of
naturally existing ones is unexpected.

Perhaps the most important piece of recent evidence in cosmology is the detection by the COBE satellite of variations in the cosmic background radiation,
whose discovery disqualified the steady-state theory. As important to cosmology as
the large-scale homogeneity of the universe is its local inhomogeneity. This has
been understood to require that the residual radiation not be completely uniform.
It must be nearly uniform if it is to be ascribed to the origination of the universe
rather, than, say, to the formation of our own galaxy or some other relatively
local, and therefore recent, phenomenon. But if the background radiation were

completely uniform, it would be a mystery how the energy of the present universe can be concentrated in discrete clumps of matter, like stars or galaxies. While quantum processes are invoked to account for local inhomogeneity, the incorporation of gravity into a unified quantum theory of forces is not addressed by variation of the background radiation. The uncertainty principles may be enough of an explanation of slight disparities in the rate of inflation, producing local inhomogeneity. The discovery of variation does not attest to any grander unification.

Chirality, an asymmetry in the weak force, is another pertinent empirical discovery. It significantly reduces the range of eligible unifying symmetries. In particular, it is the basis for selecting the process of compactification that reduces the 26-dimensional space of string theory to the 10-dimensional space of superstring theory, which appears capable of incorporating gravity without renormalization problems. But chirality is no confirmation of superstring theory, or, more generally, of the unification program, which neither explains nor predicts it. Rather, like the empirically discovered lower limit on proton decay, chirality functions as an additional constraint on the form of eligible symmetries.

The anomalous results of neutrino-detection experiments suggest another possible test of grand unification, but, unfortunately, one no more decisive than proton decay. Neutrinos captured from the sun's interior are too few by one third, compared with prediction. It is speculated that the missing neutrinos are present, but not in the form of electron neutrinos which the experiments detect. Their transmutation would confirm the violation of conservation of lepton number required by grand unified theories. Whether they could be detected in a different form depends, however, on the mass they acquire in this form. Their detection would be evidence for unified theory, but the failure to detect them can be interpreted as an indication that their mass is too small, rather that blamed on theory.

The situation depicted by this brief review of the status of relevant evidence recalls a familiar theme in discussions of theory testing. All the major theorists of scientific change distinguish individual theories from larger, extended theoretical programs to which they belong.[1] Empirical refutation of a theory need not challenge the larger program, but may actually promote it by indicating a better way to carry it out. The program as a whole is not at stake in tests of its individual

---

1. Kuhn distinguishes theories from paradigms; Lakatos, from research programs; Laudan, from research traditions. Of course, their accounts of these larger units of analysis differ, reflecting their differing views of how science changes. They share in common an opposition to the holistic view of testing advanced by Pierre Duhem and W. V. Quine, who restrict the import of evidence to large complexes of theories and auxiliary assumptions. For the theorists of scientific change, empirical evidence strikes immediately at the fortunes of localized pieces of theory, and indirectly, if at all, at extended theoretical programs. For the holists, empirical evidence is incapable in principle of discriminating among the local pieces of a larger theoretical program.

This difference is obscured in representing Kuhn as a holist, as Laudan does, for example, in *Science and Values*. Kuhn is holistic in assuming that science changes along many dimensions at once—paradigm, methodology, and cognitive goals change together. He is not a holist about testing. Indeed, since paradigms and their associated methodologies are assumed to change together, paradigms themselves, Kuhn's larger theoretical programs, become untestable. The question of the autonomy of theories and methodologies will be important in the third section of this chapter.

instantiations. This picture produces philosophical puzzles about how the more basic theoretical commitments that define a program can be objective, and about how the resolution of conflicts among programs can be rational.

Since this theme has been recalled, we must disassociate the present situation from it. The problem of objectivity may be more the creation of a philosophical picture than a question about the science pictured. The reason we cannot test the overall program of unification is not that its empirical reach is limited to that of dispensable, candidate theories. The reason is not the resilience of the unification program in the face of possible evidence. I have emphasized these conditions, but assimilating my point to that pushed by theorists of scientific change *over*emphasizes them. For there are possible empirical phenomena that would test the larger unification program directly.

The problem is that these phenomena are not supposed to occur under producible or accessible conditions. The production of X particles would confirm grand unification, for example, just as the production of the W and Z particles confirmed electroweak unification. And the failure to find the X particle under the right conditions would disconfirm grand unification, as failure to identify the W and Z would have disconfirmed the Weinberg-Salam theory and failure to find the top quark would have threatened the standard model. The trouble is that the X is so massive (10^{14} proton masses) that the right conditions require unobtainable energies. The connection between the mass of a force-carrying particle and its range (10^{-29} cm for the X) is fundamental to the quantum mechanical understanding of the forces being unified, and the connection between mass and energy is fundamental to relativity theory. These constraints are not adjustable. They at once commit the program of unification to specific empirical effects and preclude the observability of these effects.

The latter point is most important. The problem of testing the unification program has nothing to do with the methodology of theory appraisal or with alleged limitations on scientific objectivity. It is a consequence of specific claims that this program makes about the natural world, claims that might very well be false. This is unlike the reasons philosophers have given for questioning the justifiability of theoretical beliefs, and unlike the obstacles that theoretical beliefs have faced in past science. My own defense of realism is impotent in the face of this development. If this defense is the last word on the realist view of science—if, that is, novel success is the only basis for realist commitments (whether or not it is a basis at all) and my conditions for novelty are not only sufficient but necessary (which I have not claimed)—then the most fundamental theories of physics are not to be believed. Pragmatism, or less, becomes the right philosophy, for purely scientific reasons. But is mine the only route to realism?

Methodological Practice

I am inclined to think that mine is the only passable route, and, accordingly, to elevate the importance of my argument over that of the epistemology that it serves. But before acquiescing in this checkered denouement, I wish to consider another possibility. Theoretical physicists, at least those working in elementary particle theory or, more specifically, quantum gravity, are unifiers. They have doubts about superstring theory and particular GUT symmetries, but they do not doubt that the forces of nature instantiate some unifying symmetry. They do not doubt that gravity can be quantized and unified with the other forces. They believe that matter is unstable, and that under conditions that prevailed in the early universe elementary particles lose their distinct identities. Why do they believe such things, if the grounds for theoretical belief are what I have claimed? What do *they* think entitles them to such beliefs, or, if the presupposition of that question is too abstract, what *makes* them so credulous? Manifestly, my methodology is not theirs, at least not any more. What is theirs, and where did they get it?

The answers, I think, are clear enough, if difficult for philosophers to swallow. Their reasoning is straightforwardly explanationist, and they are relying on such reasoning because it is the strongest form of support available to them. In effect, they are desperate, and, in their desperation, are resorting to a standard of justification that they would much prefer to be at liberty to reject. Certainly, they want to judge theories by novel results, or, at least, more loosely, by "independent tests." This is how they got where they are, by being suspicious of speculation and hard-nosed about the dispositions of nature. Having been burned in the nineteenth century by the precipitous credulity of the method of hypothesis, scientists developed, early in the twentieth century, a healthy insistence on the empirical grounding—especially on the falsifiability—of theories, even to the point of positivism's semantic reduction of theory to observation. Surrounded by a culture whose operant mechanism of belief formation seems to be wishful thinking, a culture awash in tales of ESP and UFOs, a culture of psychic readings and horoscopes, a culture that cannot tell science from superstition, a culture intellectually as well as morally corrupted by religion, physicists see, all too plainly, the dangers of weakening their allegiance to empiricism. But their only alternative is to circumscribe the aspirations of their calling, and this evidently strikes many of them as worse.

It seems that physicists are willing to defer to their best current theories for a determination of the nature of the evidence by which current theories should be judged. Unlike social scientists, who measure their theories against a preconceived philosophical agenda modeled on what they take to be "scientific method," physicists act as though they subscribe to the following principle: *The evidence that it is appropriate to require for the acceptance of a new scientific theory is the strongest evidence that can foreseeably be obtained on the assumption that the theory is correct.* As long as a new theory fits the prevailing standards of theorizing—it has an appropriate mathematical formulation; it addresses problems and promotes desiderata that established theories identify as important; it is consistent with the established body of scientific knowledge or, where inconsistent, rationalizes its departures in ways that prevailing theories recognize as legitimate—it is not to be ruled inadmissible for failure to be amenable to existing standards of empirical evaluation. If a new theory's empirical claims, in conjunction with established background theory, render existing standards of evaluation inapplicable, then these standards will adjust sufficiently to provide for the possibility of a favorable evaluation. In short, the new theory will be given at least a fighting chance.

Of course, it must also be possible for the new theory to fail by the adjusted standards; theories are not allowed to be self-certifying. The new conditions of acceptability must be conditions that the theory could fail to satisfy. But insistence on conditions that the theory could not satisfy, whether or not it is true, is simply too restrictive. Standards of evaluation are responsive to the content of the theories required to meet them. That this is possible without inconsistency, without abandoning objectivity, is crucial to progress.

With the advent of the unificationist program in fundamental physics, we are witnessing changes of evaluative standards that elevate explanationist desiderata over novel predictive success. What is demanded of a unifying theory is not that gravitons or magnetons be discovered, but that the theory provide solutions to certain outstanding problems created, but not solved, by the more limited theories that empirical evidence has already confirmed. To be successful, a further step in unification must explain—must provide a theoretical basis for—certain hypotheses that accepted theories artificially assume, or take over from the results of experiment. Beyond that, it must reveal new connections among known phenomena and extend the applicability of established physical principles. It must forge new and unexpected connections with the rest of physics. And solutions and explanations already achieved by empirically confirmed theories must be recoverable from it. The new theory may also be required to yield novel predictions of its own. If it does all this, it may be accepted even if its own novel predictions are untestable. This is the set of requirements that seems to be in force as physicists judge the latest theoretical proposals. It is by them that certain symmetries are discounted while others are found promising, in the absence of experimental evidence capable of discriminating among them.

Although "explanationist," the standard I am describing is far more sophisticated and stringent than what is known in philosophy as "inference to the best explanation." It is insufficient that an explanation be the best; it must be *good*, which the best available explanation need not be. It is thought better to accept no theory than to accept one that leaves important problems outstanding. Of course, a problem may be obviated, rather than solved, by challenging assumptions needed to raise it. But the extant tasks that a correct theory would be expected to perform must all be addressed in a well-motivated way. These problems and tasks are quite specific, and generally independent of the theory they are used to judge.

For example, the success of the standard model for elementary particles, and of electro-weak unification, creates a number of expectations for deeper, unifying theories. There should be a theoretical basis for the number of (remaining) forces; for their relative strengths; for the different energies at which they become distinct; for the family structure of elementary particles; for the fact that certain particles are distinguishable only by their mass; for the actual masses they are found to have empirically; for the strengths of their electric charges. Constants of nature that currently appear independent and arbitrary should be connected and their relative fundamentality revealed. In particular, the vanishing smallness of the cosmological constant must be reconciled with the apparent independence of the factors that contribute to the energy density of the vacuum. Physical significance must be attached to the process of renormalization; we must discover why apparently meaningless infinities appear in solving the equations of present theories, not merely manage to evade them. By the same token, we must solve, rather than finesse, the problem of singularities in general relativity. This is an "internal problem" of relativity, comparable to the problem of infinities in quantum theory, which the quantization of gravity is expected to obviate. The importance and scope of basic principles—such as the uncertainty relations, the principle of least action, Mach's principle, and the second law of thermodynamics—must be clarified. The process of compactification and the apparent dimensionality of space-time must be explained; anthropic coincidences must be given a deeper foundation; chirality must be related to CTP conservation, and the status of partial conservation laws resolved.

These are explanatory requirements. Beyond them, a certain expectation for the "uniqueness" or "inevitability" of an acceptable theory has become important. In the absence of experimental test, theories are expected to be self-recommending, not only by providing explanations but also by precluding potential opposition, in a number of ways. One way appeals to the unified character of the corpus of established theories. There are so many connections in physics, so many divergent roles for basic principles through which they transcend the specific theory in which they originate, that an effective test of a new theory is its cohesion with the rest of physics. Many otherwise plausible avenues for meeting outstanding explanatory tasks quickly fail this test. For example, it is a remarkable feature of quantum mechanics, not present in earlier theories, that seemingly small variations in it produce conflicts elsewhere. Stephen Weinberg proposed, as an alternative to quantum mechanics, a nonlinear quantum theory that, while it made little testable difference, could not be reconciled with relativity.[2] Unlike the kind of nonlocality tolerated in quantum theory, Weinberg's theory violated special relativity's restriction on the velocity of transmission of signals. This was enough to disqualify it.

New theories proposed to unify the fundamental forces face general mathematical constraints that emerge from the success of quantum mechanics and relativity; their force laws must be both covariant and quantizable. In superstring theory, it seems possible that these constraints will suffice to identify the right equations to govern the behavior of strings. It seems possible in string theory that the correct symmetries with which to unify the forces will be dictated by the underlying mathematics, and will not have to be chosen independently and then tested for conformity with other desiderata. If so, an explanatorily successful superstring theory could claim a form of uniqueness. Empirical constraints like chirality also operate to limit the choice of symmetries used to achieve unification. Anthropic coincidences provide yet another basis for uniqueness; only theories that provide for the highly specialized conditions necessary for the evolution of intelligent life

2. *Dreams of a Final Theory*, p. 88.

are admissible. If the universe must contain observers as a condition of its very existence, as some approaches to the measurement problem of quantum mechanics suggest, then anthropic requirements appear less artificial and more as theoretically motivated constraints on new theories. Many such considerations conspire to create the sense that the search for a unifying theory is so constrained that any successful theory will not leave room for rivals.

Of course, this kind of reasoning is high qualified. Some of the constraints could be wrong, and further theorizing could supplant them. But the constraints are stronger than aesthetic or methodological preferences. The empirical success of theories that establish them argues that they are fundamental, even if the requirement of novel predictive success cannot ordain them. They are a hard-won and powerful basis for pressing theory further. A unifying theory that met our constraints, achieved our explanatory goals, and made further contributions that we do not yet know enough to demand, would not, on that account, have to be believed. Nor would we have to judge it correct to explain how it managed to do all these things. Our epistemic situation with respect to such a theory would be different from that on which I have based a case for realism. To treat such achievements as warrant for belief requires a different methodology. I suggest that the advent of such a methodology is already upon us, that we must acknowledge it to make sense of the current evaluative practice of theoretical physics. The question for the future of realism is, Can it be warranted?

Warranting Methodological Change

If we separate physical theories and the methodological standards for their evaluation into different conceptual categories, there would seem to be a logical fallacy in accepting a theory whose defense requires a change in standards.[3] To resort to changing the standards, when faced with a theory unacceptable by present standards, would seem the death of objectivity. Can one rationally prefer the theory over the standards by which theories are supposed to be preferred?

It seems clear that an affirmative answer must be possible if theoretical change is to be understandable as rational and progressive. For the fact is that methodology does change, and the source of its change must be in the development of new theories whose achievements set new standards for subsequent theory evaluation. A fully naturalistic approach to science must reject the strict bifurcation of theories and standards in favor of a dynamic in which what we expect from theories as a condition of their acceptability, and what our best theories tell us about the world, are mutually reinforcing. For only on the basis of our understanding

3. Such has been alleged by John Worrall in a review of Laudan's *Science and Values*. Laudan argues that changes of either theory or methodology can be rationalized against the stable background of the other. In particular, Kuhn's relativistic view of change depends on assuming that the two must change *ensemble*. Worrall contends that allowing methodology to change independently of theory produces a logical circularity. See "The Value of a Fixed Methodology," *British Journal for the Philosophy of Science*, 39 (1988):263–275.

of nature can we judge what it is reasonable to expect of theories. As a naturalistic account forswears any a priori basis for such expectations, it anticipates their change as our understanding of the natural world develops. Naturalism would have us examine the process by which theories and methods influence one another, and extract an abstract schema for change with which to answer the a priorist charge of circularity.

Here is mine. Let M be the prevailing methodology in terms of which currently accepted theories are justified. Let T be a new theory containing hypotheses H_i. Most, but not all, of the H_i are justified in accordance with M. Let H_n be one that is not so justified. Because of the preponderance of justified hypotheses in T, T, as a whole, is judged acceptable. (No theory is perfect, and, in practice, acceptance does not represent unqualified endorsement, nor the conviction that a theory is final or irreplaceable; such an attitude is not even part of a realist interpretation of theory, as we have seen.) H_n then gains in importance. It suggests further directions of research that prove productive as judged by M. Certain outstanding problems, not originally connected with T, problems whose solution was not considered a condition of adequacy for T, prove tractable on the basis of H_n. As a result, T comes to be valued more for H_n than for those of its hypotheses that originally passed muster by M. Rather than a defect that mars the acceptability of T, H_n comes to be regarded as an asset. The features of H_n that originally disqualified it become acceptable in further theorizing. A new methodology M', legitimating those features and retracting provisions of M that disqualify those features, supplants M.

In this schema, theories are not depicted as indivisible wholes that are accepted or rejected in toto. They are subject to internal development, and are reidentifiable through changes of content. Not all of the propositions that comprise them are *basic*, in the terminology of chapter 3. One and the same theory can continue to be judged favorably while the reasons for this judgment shift. This is the way things actually work, and only the legacy of a positivistic philosophy that attends to the finished scientific product, ignoring the dimension of theory construction (or discovery), would have us think otherwise. But for this dynamic scenario, a schema for rational, warranted change of evaluative methods does indeed run afoul of logical constraints. It can only have been an ahistorical, antinaturalistic positivism that made these constraints seem insuperable.

No doubt, the most studied, and debated, historical case of methodological change is that from Newtonian "induction from the phenomena" to the method of hypothesis. Surely, the pivotal development in this transition was the changing status of the hypothesis of universal attraction. Originally something of an embarrassment that provided a focus for criticism of Newtonian theory, it became celebrated for its explanatory and predictive power. Its success provided a precedent for hypothesizing the universal ether required by the wave theory of light. During this transition, during early-nineteenth-century debates between wave and particle theorists, it was possible to disagree about what Newtonian theory was most to be valued for, without disagreeing about the success or importance of the theory itself. The judgment that the theory was successful proved more fundamental than judgments about what was the appropriate methodology by which to judge the success of theories. This is just the switch in priorities that my schema captures.

To apply the schema to the case at hand, what has to happen for the explanationist methodology I have adumbrated to succeed is that theories *already justified* by extant empiricist standards come to be valued as much, or more, for their explanationist credentials as for the experimental successes on which their acceptance was originally based. A change in what is valued then rationalizes a change in the standards of evaluation, for the methods that it is rational to employ are those found to advance the values theories are to serve. Electroweak theory, for example, may be valued more for the unification and other theoretical connections it effects—for its discover of a new symmetry, for its renormalizability, for its structural connections to quantum electrodynamics—than for its successful predictions of new particles and of violations of parity by the weak force. The former attributes, more than the search for new, measurable effects, may—indeed, already have—come to dictate the direction of further theorizing.

A Measured Retreat

Propounding such a schema, while it arguably suffices to handle the problems that methodological change poses for rationality and progress, does not, alas, suffice to allay concerns about realism. For there are many respects in which theoretical change can be progressive that do not provide warrant for theoretical beliefs. I fear that warrant for theoretical beliefs depends on the conditions I have proposed, conditions whose epistemic import is not subsumable under a fully naturalized epistemology.

I insist on the testability of my proposal, and have argued that it meets the standard that it itself imposes on the assessment of theories. My philosophical case for realism is itself a theory, and is not exempt from the standards for judging theories. I recognize no division, in principle, between scientific theories and philosophical ones, but regard the epistemology of science as continuous with science. Further, I locate the standard of novel predictive success in the evaluative practice of at least a historical period of science. The epistemic value of novelty, which, in chapter 5, I used naturalism to defend, survives the tenure of novelty as a standard of warrant. No current or projectable evaluative practice questions the special probative weight that novel success, if achievable, would have; the question is only of its necessity. It is part of practice to prefer not to loosen evaluative standards, and to continue to seek opportunities to meet the standard of novelty. In all these respects, my position is naturalistic.

But naturalism gives me no basis for ordaining this standard for all time; indeed, if the argument of the present chapter is correct, this standard is temporary. And I do not see the grounds for maintaining that the explanationist standard that fundamental physics is coming to embrace, for all that it preserves of the rationality and progressiveness of theory preferences under a naturalistically inspired schema of change, supports a realist over a pragmatist interpretation of these preferences.

If this is a retreat, however, it is a measured one in relation to the contemporary philosophical scene. I have proposed an empirical basis for justifying theoretical beliefs. If some beliefs at the most fundamental level are not thereby justifiable, many are, at least in principle. And this result clearly exceeds the strictures of some of the more influential current epistemologies of science.

Bibliography

Achinstein, Peter. (1991). *Particles and Waves*. Oxford: Oxford University Press.

Brush, Stephen. (1989). "Prediction and Theory Evaluation: The Case of Light Bending." *Science,* 246: 1124–1129.

———. (1992). "How Cosmology Became a Science." *Scientific American,* 267: 62–70.

Cantor, G. N. (1983). *Optics after Newton*. Manchester, U.K.: Manchester University Press.

Donnellan, Keith. (1966). "Reference and Definite Descriptions." *Philosophical Review,* 75: 281–304.

Duhem, Pierre. (1969). *To Save the Phenomena*. Chicago: University of Chicago Press.

Earman, John; Glymour, Clark; and Stachel, John (eds.). (1977). *Foundations of Space-Time Theories*. Minneapolis: University of Minnesota Press.

Earman, John. (1993). "Underdetermination, Realism, and Reason." In Peter French, Theodore Uehling, and Howard Wettstein (eds.), *Midwest Studies in Philosophy,* vol. 18. Notre Dame, Ind.: University of Notre Dame Press.

Einstein, Albert. (1952). *The Principle of Relativity*. New York: Dover.

———. (1956). *Investigations on the Theory of the Brownian Movement*. New York: Dover.

Ellis, Brian. (1979). *Rational Belief Systems*. Oxford: Basil Blackwell.

Fine, Arthur. (1984). "The Natural Ontological Attitude." In Jarrett Leplin (ed.), *Scientific Realism*. Berkeley: University of California Press.

———. (1984). "And Not Antirealism Either." *Nous,* 18: 51–65.

———. (1986). "Unnatural Attitudes: Realist and Instrumentalist Attachments to Science." *Mind,* 95: 149–179.

———. (1986). *The Shaky Game: Einstein, Realism, and the Quantum Theory*. Chicago: University of Chicago Press.

Fresnel, Augustin. (1900). "Memoir on the Diffraction of Light." In Henry Crew (ed.), *The Wave Theory of Light*. New York: American Book Co.

Gardner, Michael. (1982). "Predicting Novel Facts." *British Journal for the Philosophy of Science*, 33: 1–15.

Giere, Ronald. (1983). "Testing Theoretical Hypotheses." In John Earman (ed.), *Testing Scientific Theories*. Minnesota Studies in the Philosophy of Science, vol. 10. Minneapolis: University of Minnesota Press.

Glymour, Clark. (1980). *Theory and Evidence*. Princeton: Princeton University Press.

Goodman, Nelson. (1965). *Fact, Fiction, and Forecast*. Indianapolis: Bobbs-Merrill.

Grünbaum, Adolf. (1964). "The Bearing of Philosophy on the History of Science." *Science*, 143: 1406–1412.

Hacking, Ian. (1979). "Imre Lakatos's Philosophy of Science." *British Journal for the Philosophy of Science*, 30: 381–402.

———. (1983). *Representing and Intervening*. Cambridge: Cambridge University Press.

Hartle, James B., and Hawking, Stephen W. (1983). "Wave Function of the Universe." *Physical Review*, D28, 2960–2975.

Hawking, Stephen. (1933). *Black Holes and Baby Universes*. New York: Bantam Books.

———. (1984). "The Edge of Space-Time." *New Scientist*, August 16, 1984: 10–14.

———. (1988). *A Brief History of Time*. New York: Bantam Books.

Hempel, Carl G. (1965). *Aspects of Scientific Explanation*. New York: Macmillan, Free Press.

———, and Oppenheim, Paul. (1948). "Studies in the Logic of Explanation." *Philosophy of Science*, 15: 135–175.

Hoefer, Carl, and Rosenberg, Alexander. (1994). "Empirical Equivalence, Underdetermination, and Systems of the World." *Philosophy of Science*, 61: 592–608.

Howson, Colin, and Urbach, Peter. (1989). *Scientific Reasoning*. La Salle, Ill.: Open Court.

Kitcher, Philip. (1993). *The Advancement of Science*. New York: Oxford University Press.

Kripke, Saul. (1977). "Speaker's Reference and Semantic Reference." In Peter French, Theodore Uehling, and Howard Wettstein (eds.), *Contemporary Perspectives in the Philosophy of Language*. Minneapolis: University of Minnesota Press.

Kuhn, Thomas. (1962). *The Structure of Scientific Revolutions*. Chicago: University of Chicago Press.

Kukla, André. (1993). "Laudan, Leplin, empirical equivalence and underdetermination." *Analysis*, 53: 1–7.

———. (1996). "Does Every Theory Have Empirically Equivalent Rivals?" *Erkenntnis*, 44: 137–166.

Lakatos, Imre. (1978). "Falsification and the Methodology of Scientific Research Programs." In Imre Lakatos, *Philosophical Papers*, vol. 1. Cambridge: Cambridge University Press.

Latour, Bruno, and Woolgar, Steve. (1986). *Laboratory Life*. Princeton: Princeton University Press.

Laudan, Larry. (1977). *Progress and its Problems*. Berkeley: University of California Press.

———. (1984). "Explaining the Success of Science: Beyond Epistemic Realism and Relativism." In James Cushing, C. F. Delaney, and Gary Gutting (eds.), *Science and Reality*. Notre Dame: University of Notre Dame Press.

———. (1990). "Normative Versions of Naturalized Epistemology." *Philosophy of Science*, 57: 44–60.

———. (1990). *Science and Relativism*. Chicago: University of Chicago Press.

————, and Leplin, Jarrett. (1991). "Empirical Equivalence and Underdetermination." *Journal of Philosophy*, 88: 449–472.

Laymon, Ronald. (1980). "Independent Testability: The Michelson-Morley and Kennedy-Thorndike Experiments." *Philosophy of Science*, 47: 1–38.

Leplin, Jarrett. (1972). "Contextual Falsification and Scientific Methodology." *Philosophy of Science*, 39: 476–491.

————. (1975). "The Concept of an *Ad Hoc* Hypothesis." *Studies in History and Philosophy of Science*, 5: 309–345.

————. (1981). "Truth and Scientific Progress." *Studies in History and Philosophy of Science*, 12: 269–293.

————. (1982). "The Assessment of Auxiliary Hypotheses." *British Journal for the Philosophy of Science*, 33: 235–249.

————. (1986). "Methodological Realism and Scientific Rationality." *Philosophy of Science*, 53: 31–51.

————. (1987). "The Bearing of Discovery on Justification." *Canadian Journal of Philosophy*, 17: 805–814.

————. (1987). "The Role of Experiment in Theory Construction." *International Studies in the Philosophy of Science*, 2: 72–83.

————. (1988). "Is Essentialism Unscientific?" *Philosophy of Science*, 55: 493–511.

————. (1988). "Surrealism." *Mind*, 97: 519–524.

————. (1994). "Critical Notice: The Advancement of Science." *Philosophy of Science*, 64: 666–671.

————, and Laudan, Larry. (1993). "Determinism Underdeterred." *Analysis*, 53: 8–16.

Levin, Michael. (1984). "What Kind of Explanation Is Truth?" In Jarrett Leplin (ed.), *Scientific Realism*. Berkeley: University of California Press.

Lorentz, H. A. (1915). *The Theory of Electrons*. New York: G. E. Stechert.

Lycan, William. (1988). *Judgment and Justification*. Cambridge: Cambridge University Press.

Mayo, Deborah. (1991). "Novel Evidence and Severe Tests." *Philosophy of Science*, 58: 523–553.

Mettlin, C. (1991). "Vasectomy and Prostate Cancer Risk." *American Journal of Epidemiology*, 7: 107–109.

Michelson, A. A., and Morley, E. W. (1887). "On the Relative Motion of the Earth and the Luminiferous Aether." *Philosophical Magazine*, 5: 449.

Miller, Richard. (1989). "In Search of Einstein's Legacy." *Philosophical Review*, 98: 215–239.

Musgrave, Alan. (1974). "Logical Versus Historical Theories of Confirmation." *British Journal for the Philosophy of Science*, 25: 1–23.

Nagel, Ernest. (1961). *The Structure of Science*. New York: Harcourt, Brace, and World.

Norton, John. (1995). "Eliminative Induction as a Method of Discovery: How Einstein Discovered General Relativity." In Jarrett Leplin (ed.), *The Creation of Ideas in Physics*. Dordrecht: Kluwer Academic Publishers.

Page, Don. (1995). "The Hartle-Hawking Proposal for the Quantum State of the Universe." In Jarrett Leplin (ed.), *The Creation of Ideas in Physics*. Dordrecht: Kluwer Academic Publishers.

Putnam, Hilary. (1981). *Reason, Truth, and History*. Cambridge: Cambridge University Press.

————. (1984). "What Is Realism?" In Jarrett Leplin (ed.), *Scientific Realism*. Berkeley: University of California Press.

Salmon, Merrilee, et al. (1992). *Introduction to the Philosophy of Science.* Englewood Cliffs, N. J.: Prentice Hall.

Schilpp, P. A. (1959). *Albert Einstein: Philosopher-Scientist.* New York: Harper Torchbooks.

Sellars, Wilfrid. (1961). "The Language of Theories." In Herbert Feigl and Grover Maxwell (eds.), *Current Issues in the Philosophy of Science.* New York: Holt, Rinehart, and Winston.

Shankland, R. S. (1963). "Conversations with Einstein." *American Journal of Physics,* 31: 47–57.

Stachel, John. (1995). "Scientific Theories as Historical Artifacts." In Jarrett Leplin (ed.), *The Creation of Ideas in Physics.* Dordrecht: Kluwer Academic Publishers.

Stove, David. (1991). *The Plato Cult.* Oxford: Basil Blackwell.

Torretti, Roberto. (1990). *Creative Understanding.* Chicago: University of Chicago Press.

———. (1995). "Einstein's Luckiest Thought." In Jarrett Leplin (ed.), *The Creation of Ideas in Physics.* Dordrecht: Kluwer Academic Publishers.

Van Fraassen, Bas. (1980). *The Scientific Image.* Oxford: Oxford University Press.

———. (1985). "Empiricism in the Philosophy of Science." In M. Churchland and C. A. Hooker (eds.), *Images of Science.* Chicago: University of Chicago Press.

———. (1989). *Laws and Symmetry.* Oxford: Clarendon Press.

Weinberg, Stephen. (1992). *Dreams of a Final Theory.* New York: Vintage Books.

Whewell, William. (1840). *The Philosophy of the Inductive Sciences.* London: J. W. Parker; New York: Johnson Reprint Corp., 1967.

Whittaker, E. T. (1951). *A History of the Theories of Aether and Electricity.* London: Thomas Nelson and Sons.

Worrall, John. (1984). "An Unreal Image." *British Journal for the Philosophy of Science,* 35: 65–80.

———. (1985). "Scientific Discovery and Theory Confirmation." In Joseph C. Pitt (ed.), *Change and Progress in Modern Science.* Dordrecht: D. Reidel.

———. (1988). "The Value of a Fixed Methodology." *British Journal for the Philosophy of Science,* 39: 263–275.

———. (1989). "Fresnel, Poisson and the White Spot: The Role of Successful Predictions in the Acceptance of Scientific Theories." In D. Gooding, T. Pinch, and S. Schaffer (eds.), *The Uses of Experiment.* Cambridge: Cambridge University Press.

Zahar, Elie. (1973). "Why Did Einstein's Programme Supersede Lorentz's?" *British Journal for the Philosophy of Science,* 24: 95–123.

Index

abduction, 22, 82, 114–118, 132–
    133, 138, 165n, 166, 168, 174–
    176. *See also* reasoning
aberration, astronomical, 88–89
Achinstein, Peter, 37n
adequacy
    empirical, 22–23, 28, 138, 161n,
        164n, 165–168, 172, 177
    minimal, 71–72
    of reconstructions, 60–71
    See also reconstructions; theories
aesthetics, 57, 126
Airy, George, 36, 74, 89
algorithms, 13, 153, 158–161
Arago, François, 59–60, 62, 73, 85,
    89
Aristotle, 27n, 135n
arithmetic, 67, 173
astronomy, 69, 105
    ancient, 146
    Copernican, 42n, 43, 105n
    Ptolemaic, 42n, 105n

    See also cosmology
auxiliaries. *See* hypotheses

Balmer series, 42
Bayes's theorem, 44–48, 61. *See also*
        confirmation; probability
belief(s), 46–47, 122, 163, 166–168,
        179
    empirical, 5, 164–165
    perceptual, 152–153, 159, 169
    scientific, 4, 174
    theoretical, 170–171, 176, 182–
        183, 185, 188–189
Berkeley, George, 143
bias, 3–5, 8
biology, 110–111
    evolutionary, 8, 11, 35
    molecular, 8, 12
Biot, Jean Baptiste, 36–37
Bohr, Niels, 16, 42, 175
Borg, Bjorn, 9
Brahe, Tycho, 6

Brown, Robert, 75
Brownian motion, 75, 99
Brush, Stephen, 99n

Cantor, C. N., 84n
causation, 107, 110, 118
commitment(s), 22, 24–25, 28, 38,
    40–41, 102, 152–153, 163, 182
  empirical, 154–155, 168
  epistemic, 132, 138, 166, 177
  ontological, 18
  pragmatic, 132 (*see also*
    pragmatism)
  theoretical, 67, 72, 138, 158, 169,
    180, 182
confirmation, 12–13, 34–35, 37–38,
    45, 56, 62, 73, 98
  Bayesian, 37n, 44–48, 108–109,
    114n, 128
  equivalence condition of, 110–111
  transitivity of, 121
constructivism. *See* mathematics
contraction
  effect, 89
  hypothesis, 76, 92
  See also ether, hypothesis of;
    Lorentz, H. A.
Copernicus, Nicholas, 73, 105
cosmological constant, 76, 77n, 93
cosmology, 70, 76, 77n, 153n
  big bang, 94–97, 131n, 138, 144,
    149–150, 174
  inflationary, 179–181
  steady-state, 96–97, 138, 174, 180
cumulativity, 137, 146–149, 151–152

Darwin, Charles, 35–36
deduction, 8, 16–17, 65, 70, 110,
    174. *See also* method(s)
definition(s), 67–68, 71. *See also*
    meaning(s)
DeLisle, J. N., 59n
Descartes, René, 19, 35n, 134n, 159
Dickie, Robert, 97
diffraction, 29, 36–37, 58–59, 63, 73,
    83–86, 146–147

discovery, 44, 187
  contrasted with justification, 48,
    65, 86
  See also theories
Duhem, Pierre, 35n, 38, 155, 161–
    162, 181n

Earman, John, 108n
Eddington, Arthur, 73–74, 99n, 130
EE (empirical equivalence thesis),
    137, 152–163, 165. *See also*
    equivalence
effects, scientific, 10, 72–74. *See also*
    contraction; experiment(s);
    novelty; photoelectric effect
Einstein, Albert, 10–11, 30–31, 46,
    49, 51–52, 70, 75–76, 79, 86–
    96, 99, 101n, 106n, 122n, 123–
    124, 129, 135n, 175. *See also*
    relativity
electromagnetism, 16, 30, 43, 69, 79,
    87, 90, 92, 131. *See also*
    Maxwell's equations
electroweak theory, 70–71, 125, 179,
    182, 184, 188. *See also* forces of
    nature
Ellis, Brian, 28
empirical consequence classes of
    theories. *See* theories,
    observational consequences of
empiricism, xi, 46, 117, 169, 183,
    188
  constructive, 164–166, 168–172,
    176
energy, 91–93
  of blackbody radiation, 94, 97
  kinetic, 10, 91–92
  potential, 18
  quantized, 93
  See also cosmology; forces of
    nature; quantum mechanics
epistemology, xi, 19–21, 40, 43, 47,
    60n, 99, 135, 162–163, 179,
    182, 189
  evolutionary, 7

naturalistic, 82, 102, 119, 132n, 145–146, 175, 186–188
probabilistic, 45–46, 123n, 165n
equivalence
condition (*see* confirmation)
empirical, 80, 152, 164n, 168, 177 (*see also* EE)
of mass and energy (*see* relativity)
mathematical (*see* mathematics)
principle of, 70, 79, 94, 106, 122n
See also EE; Einstein, Albert; Galileo Galilei; relativity
ether, hypothesis of, 18, 36, 69, 76, 83, 89, 130–131, 142, 146–148, 187. *See also* electromagnetism; light; Michelson-Morley experiment; relativity
evidence, 35, 37, 41n, 45–47, 55–58, 67, 96, 100, 105, 107, 109, 113, 122, 128, 137, 154–155, 160, 163, 165, 177, 179, 181n, 183
limitations of, 13, 19, 22, 24, 32
problem of old, 45
See also tests, severity of; theories
evolution, 29, 32, 35
of the universe, 94, 97, 150, 161, 179
See also biology
experiment(s), 59–60, 85, 89
controls on, 8, 30, 111
first-order in *v/c*, 70, 87, 89–91
individuation of, 39–40, 106
thought, 87
See also Michelson-Morley experiment
explanation, 7–15, 18–25, 28, 31, 35, 38, 118, 165n, 175, 185, 188
deductive-nomological model of, 20–21
evolutionary, 8, 11, 12, 23
pragmatic analysis of, 11, 23, 138, 172
realist, 25n, 27, 62, 100, 121, 151–52, 172, 183
of regularities, 81, 111–116
requirement for novelty, 63, 78
See also prediction; realism; success
explanationism. *See* abduction; MER; realism

failure of theories, 6–8, 12, 16, 32–33, 105, 136, 143–145, 149, 168. *See also* theories
fallibilism, 39. *See also* methodology
Faraday, Michael, 124
Feynman, Richard, 149
Fine, Arthur, 21, 24–26, 133n, 137–141, 173–175
Fitzgerald, C. F., 76
Fizeau, Armand, 88–89
forces of nature, 70–71, 124–125, 170, 179
unification of, 32, 125, 179–185 (*see also* physics)
Fresnel, Augustin, 36–37, 58, 60, 62, 73, 84–86, 89, 96–97, 106, 146–148
Friedmann, Alexander, 70

Galileo Galilei, 21
laws of, 30, 43, 51, 70, 149
Gardner, Michael, 42n, 51–54, 58, 99–100
gauge invariance. *See* symmetry principles
Gell-Mann, Murray, 170
Gentzen, Gerhard, 173–174
geometry, 94, 150
Giere, Ronald, 35n, 58–63
Glymour, Clark, 45n
goals of science, 3, 100–103, 169, 172, 181n, 185
Gödel, Kurt, 173
Goodman, Nelson, 111–112
gravity, 11, 32, 38, 51, 73, 77–78, 125, 130, 134n, 149, 180–181, 183. *See also* Newtonian theory; quantum mechanics; relativity; string theory
Grimaldi, Francesco, 58n
Grünbaum, Adolf, 41

GUT (grand unified theory), 125,
    179–181, 183. *See also* forces of
    nature

Habicht, Conrad, 51n
Hacking, Ian, 10–11, 41n, 42n,
    170n, 171n
Hartle, James B., 149–150
Hawking, Stephen, 97, 131n, 149–
    150
Hempel, Carl, 20
heuristics, 7, 21, 50, 56, 71, 137
Hilbert, David, 122n, 140, 173
historians of science, 4, 6n, 122
history
    determinism in, 122
    of knowledge, xii
    of science, 6, 117, 134–136, 144
Hoefer, Carl, 161n
holism, 155–156, 161n, 162, 181n
Hooke, Robert, 35n, 83
Hopkins, William, 35
Howson, Colin, 45n
Hubble, Edwin, 93–96
Hume, David, 81–82, 107–108, 110,
    112–113, 117–118
Huygens, Christian, 35n, 58, 83
Huygens's principle, 59, 84–85
hypothesis, method of. *See* method(s)
hypotheses, 6–8, 21, 31, 34–35, 38–
    39, 130
    auxiliary, 16, 22, 38–39, 65, 74–75,
        78–79, 95–96, 129, 154–158,
        160–163, 181n
    basic, 69–71, 74, 149, 187
    testing of, 7, 61–62
    See also tests; theories
hypothetico-deduction. *See*
    method(s)

idealism, 143
idealization, 103
incommensurability, 68. *See also*
    Kuhn, Thomas
independence, 31, 34, 39, 48, 59n,
    62–63

condition for novelty, 64–65, 77,
    80, 85–86, 91, 96, 100, 123,
    151–152
    among empirical results, 40–44,
        49, 63
    of empirical results from theories,
        49–50, 58, 84
    requirement for novelty, 63
induction, 12, 38–39, 106, 117, 166–
    167
    consilience of, 39–41
    eliminative, 113–115, 146n, 163
    enumerative (straight), 81–82,
        106–116, 164n, 174–176
    mathematical, 173
    new riddle of, 111–113
    paradoxes of, 81, 110–113, 143
    problem of, 81, 107–108
    skeptical historical, 135n, 136,
        139, 141–146, 148–152, 174–
        175
inertia, 92
    of energy, 73, 91
    principle of (*see* Newton's laws)
inference
    ampliative, 71, 110, 112, 115–116,
        119, 159, 165n, 166, 168, 173–
        174
    to the best explanation (*see*
        abduction)
    explanatory (*see* abduction)
    modes (forms) of, 38, 110, 112,
        114, 118, 140, 173
    See also deduction; induction;
        method(s)
inquiry, 3–5, 6n, 29, 38, 122, 132,
    139–140, 175, 179
instrumentalism, 25n, 105–106, 110,
    115, 131, 137, 139, 161n, 176
interests, 3, 11–12, 50, 71, 122, 134–
    135, 137, 139
intuition(s), 27, 98–99, 101–102

justification, 25, 40, 78, 82, 95, 107n,
    152–153, 172, 175, 182–183
    of induction, 107n

See also discovery; novelty; success; theories

Kekulé, August, 54
Kennedy-Thorndike experiment, 106
Kepler, Johannes, 6, 106
Kepler's laws, 30–31, 148–149
Keynes, John Maynard, 43
King, John, 133n
Kitcher, Philip, 146–148
knowledge, xi–xii, 10–11, 19–20, 39, 80, 99, 144, 154
  auxiliary (*see* hypotheses)
  background, 44–46, 68, 114, 116, 155, 183
  construction of, 3–5
  empirical, 5, 128, 144
  growth of, 29, 40, 100n
  scientific, 7, 28, 39, 66, 131, 174, 183
Kuhn, Thomas, 6, 41n, 67, 99, 117n, 135, 149, 181n, 186n
Kukla, André, 159–160, 174

Lakatos, Imre, 40–43, 45, 49–50, 53–54, 149, 181n
language(s), 67–68, 117
  formal, 67
  natural, 66, 112
  object versus meta-, 140
Laplace, Pierre Simon, 36
Laudan, Larry, 4n, 8–9, 13, 20–21, 80n, 132n, 137, 140n, 149, 153n, 167n, 177, 181n, 186n
law(s), 20, 24, 31–32, 87–88, 90, 92, 110, 149–150
  empirical, 15, 126
  theoretical, 17
  See also, electromagnetism; Galileo Galilei; Maxwell's equations; Newton's laws; Kepler's laws; relativity
Levin, Michael, 10–11
light
  independence of (*see* velocity of)

  gravitational deflection of, 73–74, 79–80, 99n, 130
  theories of, 16, 36–37, 46, 58, 84–85, 142, 147–148, 175, 187
  velocity of, 30–31, 37, 40, 76, 78, 86–92, 127, 129
  See also, diffraction; electromagnetism; physics; relativity
Lorentz, Henrich A., 54, 69, 88, 91–92, 123
  electron theory of, 69, 76, 123, 129
Lorentz transformations, 91–92, 124
Löwenheim-Skolem theorem, 137n
Lycan, William, 163n

mass, 51, 70, 73, 79, 91–93. *See also* quantum mechanics; relativity
mathematics, 20, 47, 69, 122n, 123, 140, 153n, 173–174, 185. *See also* arithmetic; geometry; probability; symmetry principles
Maxwell, James Clerk, 16, 69, 123–124, 126, 180
Maxwell's equations, 18, 69, 86–87, 91–92, 129
Mayo, Deborah, 55, 61–62
McEnroe, John, 9
meaning(s), 67, 117
  verificationist theory of, 104
measurement, 74, 186
mechanics. *See* Newton's laws; quantum mechanics, relativity
MER (minimal epistemic realism), 26, 102–106, 119–120, 132, 136, 141, 145–146, 148–149, 151, 153, 156, 160n, 163. *See also* realism
Mercury, precession of, 42–43, 49–52, 74, 78–80, 99n, 123. *See also* relativity
metaphysics, 19–21, 28, 31–33, 98, 139, 142, 158, 179. *See also* truth

method(s), 56, 99
    of hypothesis, xii, 34–35, 37–39,
        84, 183, 187
    hypothetico-deductive, 6–7, 12–13,
        34, 38, 65
    inductive, 35–38, 187
    of inference (*see* inference)
    scientific, 3, 5–6, 31–32, 38, 138,
        169, 171–172, 183
    See also abduction; methodology
methodology
    Bayesian (*see* confirmation)
    changes of, 37–38, 125, 141, 181n,
        184, 186–187
    debates over, 36–37, 47
    falsificationist, 38–39, 75, 174, 183
    of inquiry, 3, 99
    prescriptive, 70, 82, 88–90, 101
    of research programs, 42–43
    of theory evaluation, 21, 37, 182–
        188
    warrant for, 186–187 (*see also*
        epistemology)
    See also method(s); Mill, John
        Stuart; Whewell, William
Michelson, Albert A., 30–31, 40n,
    76
Michelson-Morley experiment, 30,
    42n, 43, 76, 86, 88–93, 123–124
Mill, John Stuart, 35–37, 41n, 43
Miller, Richard, 135n
Minkowski, Hermann, 11, 175
miracles, 7n, 100n, 172
morality, 20, 101, 107n, 143

Nagel, Ernest, 99
naturalism. *See* epistemology
Newton, Issac, 21, 31, 35, 38, 51, 58,
    77, 79, 82, 84–85, 89, 134,
    135n, 141, 148
Newtonian theory, 30, 69, 73–74,
    78–79, 87–88, 90, 134, 148,
    153n, 187
Newton's laws, 31, 36, 51, 69, 77, 90,
    106, 129, 149

no-boundary proposal, 149–150
NOA (natural ontological attitude),
    25, 133n, 139–141, 174–175
novelty, 40, 47, 49, 62–63, 72, 81,
    146
    analysis of, 43, 60, 63, 65, 74–75,
        77–78, 80, 104, 126
    of empirical results, 40n, 58, 85,
        91–94, 96–97, 132, 154, 163
    epistemic significance of, 43, 50–
        53, 57, 71, 95–99, 101–102,
        120–121, 124, 152, 165–166,
        172
    influence of, 53, 99
    knowledge criterion of, 52–54
    paradigms of, 43, 62, 96
    psychologized, 42n, 50–51
    sociologized, 51
    temporal criterion of, 41–43, 45,
        50, 54–55, 58–59, 99n, 120
    use criterion of, 50, 54–56, 58, 120
    See also independence; qualitative
        generalization; success;
        uniqueness

objectivity, 37, 67–68, 75, 96, 102,
    120–121, 182, 184, 186
observation(s), 6n, 23–24, 72, 96,
    100, 102, 117, 137, 156, 162,
    179, 183
    distinguished from theory, xi, 167
    See also theories
Ockham's razor, 25
Oddie, Graham, 48n
ontology, 11, 18, 88, 143, 170
Ostwald, Wilhelm, 75, 99
Owen, Richard, 35

parallax, stellar, 72–73
Peebles, James, 97
Penrose, Roger, 97
Penzias, Arno, 97
Perrin, Jean Baptiste, 75–76

philosophers, 6n, 10, 15–16, 31, 40,
62–63, 67, 98, 136, 183. *See
also* realists, scientific
philosophy, 99, 101, 139–140, 174–
175, 179, 182
of mind, 143, 164n
of science, 85, 99, 117, 139, 171,
174
phlogistic chemistry, 130–131, 142–
143, 146–147
photoelectric effect, 10, 29, 46
physicists, 32, 99, 135n, 170, 179,
183–184
physics, 18, 74, 77n, 117, 123, 125,
134, 141, 149, 184
contemporary, 128–129, 135, 148–
149, 161, 178, 182, 186, 188
elementary particle, 169–171, 179,
183
unity of, 185
See also astronomy; cosmology;
forces of nature; Maxwell's
equations; Newtonian theory;
quantum mechanics; relativity
Planck, Max, 93, 122
Poincaré, Henri, 54
Poisson, Siméon Denis, 36–37, 59,
85
Popper, Karl, 38–39, 46, 132n, 174
positivism, xi, 183, 187
practice, 66, 101
evaluative, 21, 102, 185, 188
historical, 77
inferential, 112
linguistic, 100
scientific, 13, 51, 53, 71, 82, 99,
101, 121, 132, 152, 169–172
pragmatism, 105, 132–133, 135, 137,
139, 176–177, 182, 188. *See also*
truth; virtues
prediction(s), 12–17, 31, 34–35, 42,
73, 81, 93, 106, 156
compared with explanation, 41n
flexibility of, 74
novelty of, 41, 75, 119, 172, 176

quantitative accuracy of, 74–76,
118n
See also novelty; success
presuppositions, 67–68, 154–156,
174
probability, 13, 37, 42, 49, 58, 61,
108, 166n, 172
personalist interpretation of, 44,
46, 109
physical, 60n
as relative frequency, 61–62
See also Bayes's theorem;
confirmation
problems, 67, 132, 184, 187
experimental, 70
historical, 52, 95, 99
philosophical, 39, 41
scientific, 7, 15, 31, 41, 49–50,
126, 134
See also novelty
progress, scientific, xi, 3, 40, 42, 67,
100n, 133, 135, 144–145, 161n,
175, 179, 184, 186, 188
projection, 81–82, 105, 111–113,
115, 118–119. *See also*
induction
proof, 20, 99, 140, 173–174
burden of, 15, 101
provenance of theories. *See* theories
psychology, 159–160, 164. *See also*
novelty
Ptolemy, Claudius, 69. *See also*
astronomy
Putnam, Hilary, 28, 100n

qualitative generalization, 73–75, 77,
91, 95, 97, 100, 127
quantum mechanics, 18, 32–33, 70,
122, 141, 144, 149, 175, 182,
184–186
indeterminacy in, 9, 12, 128, 161,
181
quark theory, 169–171, 179
Quine, W. V., 67, 80, 137, 155, 161–
162, 181n

rationality, 41, 46–47, 67–68, 103,
　132n, 168n, 174, 182, 186,
　188
realism, scientific, 26–28, 39, 63, 81,
　99–102, 110, 119–120, 122, 124,
　127, 130, 131n, 133n, 136–137,
　139–141, 143, 145–148, 151,
　153–155, 159, 162, 165–166,
　168, 170n, 172–173, 182, 185,
　187
　interpreted, 103–104
　novel predictions of, 175–177
　qualified, 97, 188–189
　types of, 26
　See also abduction; explanation;
　　MER; truth
realists, scientific, 121, 140, 167, 169.
　See also realism, scientific
reasoning, 65–66, 72, 126, 135, 140,
　163, 168, 174
　a priori, 102, 139, 174, 185, 187
　explanationist (*see* abduction;
　　explanation)
　practical, 82
　reconstruction of (*see*
　　reconstructions)
　See also abduction; adequacy;
　　deduction; induction
reasons, 66, 70, 81, 149, 170n, 171,
　182
reconstructions, 68, 70–73, 85, 91,
　123, 126–127
reference, 18, 131n, 147–148, 175
Reichenbach, Hans, 48, 65
relativism, 41n, 132n, 143, 177,
　186n
relativity
　general theory of, 11, 32, 43, 49–
　　52, 70, 73–74, 78–80, 93–97,
　　99n, 122–123, 130, 134n, 150,
　　180, 185
　in mechanics (*see* Newtonian
　　theory)
　principle of, 86–93, 129
　special theory of, 11, 18, 30, 40–
　　43, 54, 70, 73, 76, 78–79, 86–

　93, 106, 124, 127, 134, 141,
　175, 182
reliability
　of methods, 137, 139, 145
　of tests, 60–62
　of theories, 9, 27–28, 106, 109–
　　110, 115, 119, 165, 176–177
representation(s), 9, 26, 103–105. *See
　also* truth
research, 4, 7, 36, 96, 169
　programs, 40–43
revolutions, 135. *See also* Kuhn,
　　Thomas; methodology; scientific
　　change
risk, epistemic, 24, 164–169, 172
Romer, Olaus, 89
Rosenberg, Alexander, 161n
Rutherford, Ernest, 170

Salmon, Wesley, 108n
Schwartzchild approximation, 74
science, xi, 3–6, 18, 29, 99, 119, 122,
　134–135, 137, 139–142, 145,
　174–175
scientific change, 117, 132, 135, 137,
　144, 146, 149, 151, 154, 161n,
　181–182, 186, 188
scientists, 4, 36, 43, 47, 67–68, 96,
　119, 131–132, 174, 183
　community of, 36, 51, 70, 95, 129,
　　152
　political, 20
　social, 4, 5, 183
　See also physicists
Sellars, Wilfrid, 100n
severity. *See* tests
simplicity, 39, 70, 72, 87, 124–125,
　137, 166
skepticism, xi–xii, 26, 82, 115, 144,
　155, 166–167, 169. *See also*
　　Hume; induction
sociologists, 4
sociology of science, 4
Stachel, John, 124n
standard model, 169, 182, 184
Star Trek, 46n

Stove, David, 5n
string theory, 125, 181, 183, 185
subjectivity, 20, 56
success
    empirical, 15, 136–137, 185
    explanatory, 64 (*see also*
        explanation)
    novel, 42n, 62, 98–104, 118–119,
        121–122, 127–132, 146, 148–
        151, 162–163, 168, 172–173,
        175–176, 182, 184–185, 188
    predictive, 14–16, 18, 23, 31, 39,
        43–44, 48, 75, 93, 96, 119, 142,
        147, 151
    referential, 136–137 (*see also*
        reference)
    of science, 3–6, 9, 26, 30, 100,
        107n, 141
    of theories, 8–9, 12–14, 17, 20–24,
        26–27, 29, 33, 100–102, 105,
        116, 118, 129, 172–173
    See also novelty; realism
support, epistemic. *See* evidence;
        novelty; theories; success
surrealism, 26–28, 159
Svedberg, Theodor, 75
symmetry principles, 70, 124–125,
        129, 180–181, 183–184

tests, severity of, 55, 61–62. *See also*
        theories
theoretical terms, 67
theories, 5, 6n, 7–10, 13, 31, 58, 80
    acceptance of, 12, 16, 58, 60, 137,
        152, 183–184, 186–187
    changes of (*see* scientific change)
    choice of, 67–68
    construction of, 7, 19, 48–49, 55,
        58, 187
    essential (basic) propositions of,
        69, 129, 131
    of everything, 161
    formal, 66–67, 138n
    individuation of, 13, 67–69, 123,
        129

as instruments, xi (*see also*
        instrumentalism)
    of method, 6 (*see also* method(s))
    model-theoretic definition of, 22,
        66–68, 159n
    observational consequences of, 22,
        105, 130, 154–164 (*see also*
        adequacy; empiricism)
    philosophical, 98–99, 101–102
    propositional view of, 66–68, 129,
        159
    provenance of, 48, 56, 58, 66, 68,
        85, 88, 94, 96–97, 121, 123–
        124, 126–127
    reconstructions of (*see*
        reconstructions; adequacy)
    rejection of, 61, 74, 130–132, 144,
        151–151 (*see also* failure of
        theories)
    revision of, 69, 97, 129
    semantic analysis of, 22, 66–68,
        159n
    statistical, 15
    testing of, 8, 12, 16, 32, 55, 58, 60,
        68–69, 79, 155, 177, 181 (*see
        also* tests, severity of)
    total, 128, 161–163
    See also confirmation; evidence;
        success; truth
theory-ladenness, 67, 96, 117
Torretti, Roberto, 118n
truth, 19, 25, 29–32, 39–40, 174
    as accuracy of representation, 29,
        32, 103–104, 127–131, 134–135
    approximate, 15–16, 18, 103, 127,
        129–135, 137, 142, 151
    of explanations, 10–11, 20–21
    indicators, 138, 165, 169, 172
    metaphysical nature of, 29, 32–33,
        131, 134
    partial, 15–16, 18, 103, 127, 129–
        135, 137, 142, 151
    of predictions, 13–14
    reductive analysis of, 29
    redundancy theory of, 17–18
    significant, 104

truth *(continued)*
  similarity to, 15–16, 23
  surrogates for, 25, 28–29, 32
  of theories, 14–16, 18, 22–24, 27–
    28, 31, 33, 60–61, 64, 75, 100,
    105, 137, 163, 172–173 *(see also*
    MER; realism)

UD (underdetermination thesis),
  137, 153–163, 165. *See also*
  underdetermination
underdetermination, 11–15, 17, 19–
  20, 26, 28, 67, 144–145. *See*
  also UD
understanding. *See* explanation
uniqueness
  condition for novelty, 64–65, 75–
    77, 80–82, 85–86, 91, 115, 151–
    152, 176
  of fundamental theory, 128, 185
Urbach, Peter, 45n

van Fraassen, Bas, 7, 11, 21, 23–26,
  27n, 60, 137–138, 159n, 161n,
  164–169, 172–173
verisimilitude. *See* truth
virtue(s)
  pragmatic, 120, 137, 170, 177
  theoretical, 9, 138, 166, 169
von Ignatowski, W., 124n

Weinberg, Stephen, 111n, 185
Weinberg-Salam theory, 182
Whewell, William, xii, 34–36, 39–40
Whittaker, E. T., 59n
Wilson, Robert, 97
Worrall, John, 55–58, 59n, 138n,
  186n

Young, Thomas, 36, 58, 83–85

Zahar, Elie, 42–43, 49–50
Zweig, George, 170

DATE DUE

DEC 2 1 1998			

Demco, Inc. 38-293